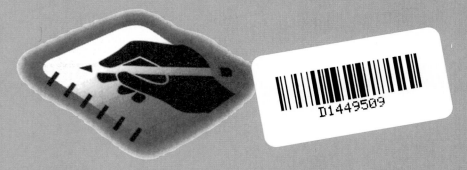

anatomy &
Anatomy & Physiology

physiology

fourth edition

Rod R. Seeley

Trent D. Stephens

Philip Tate

**WCB
McGraw-Hill**

Boston, Massachusetts   Burr Ridge, Illinois   Dubuque, Iowa
Madison, Wisconsin   New York, New York   San Francisco, California   St. Louis, Missouri

WCB/McGraw-Hill

A Division of The **McGraw·Hill** *Companies*

Student Study Art Notebook to accompany
Seeley/Stephens/Tate *Anatomy and Physiology*,4e.

 Recycled paper
This book is printed on recycled paper containing 10% postconsumer waste.

1 2 3 4 5 6 7 8 9 0 VNH VNH 9 0 9 8 7

ISBN 0-697-39479-4

TO THE STUDENTS

The *Student Study Art Notebook* is designed to help you in your study of human anatomy and physiology. The notebook contains art reproduced from the textbook. Each figure also corresponds to one of the 400 overhead transparencies; thus you can take notes during lectures, or jot down comments as you are reading through the chapters.

The notebook is perforated and 3-hole punched, so if you wish, you can remove sheets and put them in a binder with other study or lecture notes. Any blank pages at the end of this notebook can be used for additional notes or drawings.

We hope this notebook, used along with your text, helps to make the study of the human body easier for you.

DIRECTORY OF NOTEBOOK FIGURES

TO ACCOMPANY
SEELEY/STEPHENS/TATE
ANATOMY AND PHYSIOLOGY, 4E.

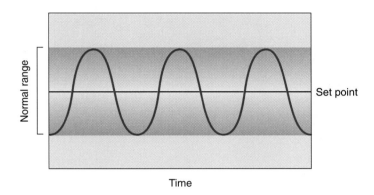

Homeostasis
Figure 1.3

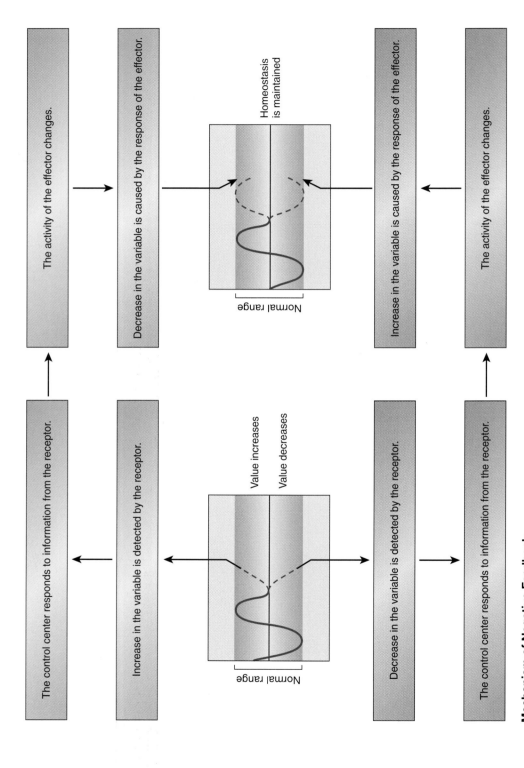

Mechanism of Negative Feedback
Figure 1.4

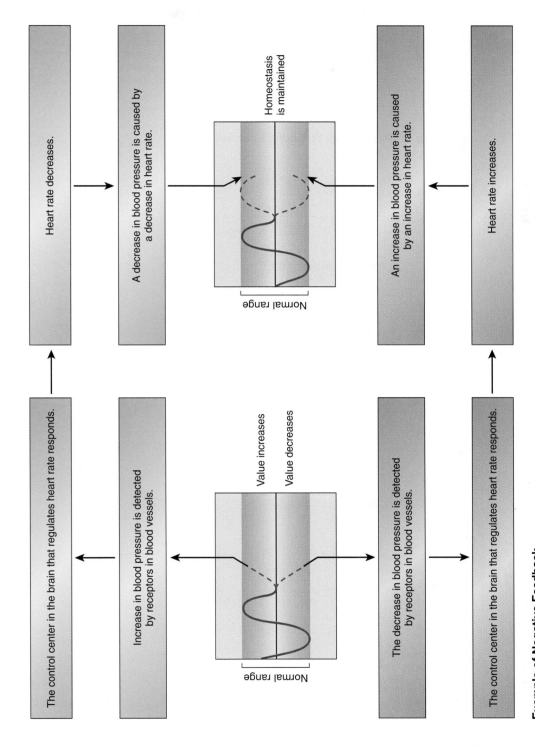

Example of Negative Feedback
Figure 1.5

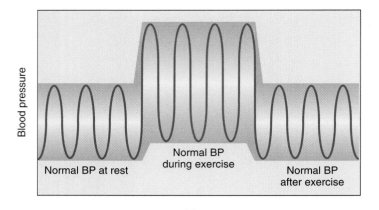

Time

Changes in Blood Pressure During Exercise
Figure 1.6

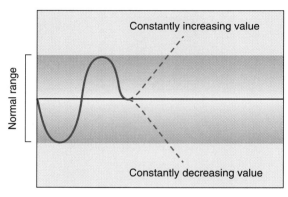

Time

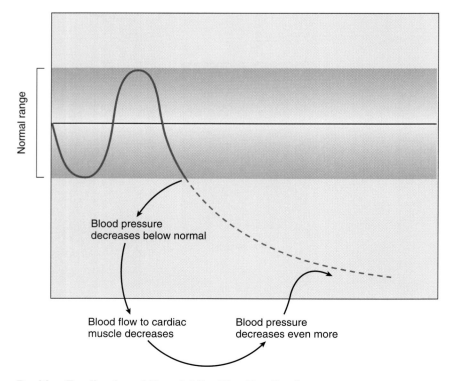

Positive Feedback and Harmful Positive Feedback
Figures 1.7, 1.8

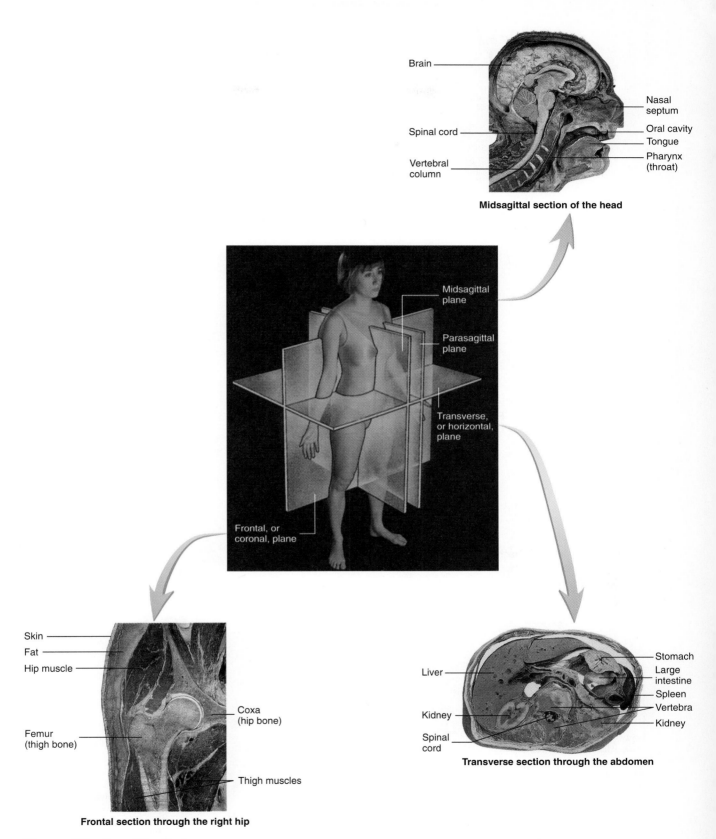

Brain

Spinal cord

Vertebral
column

Nasal
septum
Oral cavity
Tongue
Pharynx
(throat)

Midsagittal section of the head

Midsagittal
plane

Parasagittal
plane

Transverse,
or horizontal,
plane

Frontal, or
coronal, plane

Skin
Fat
Hip muscle

Femur
(thigh bone)

Coxa
(hip bone)

Thigh muscles

Frontal section through the right hip

Liver

Kidney

Spinal
cord

Stomach
Large
intestine
Spleen
Vertebra
Kidney

Transverse section through the abdomen

Planes of Section of the Body
Figure 1.10

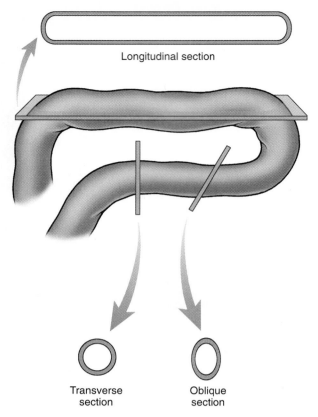

Longitudinal section

Transverse section

Oblique section

Planes of Section Through an Organ
Figure 1.11

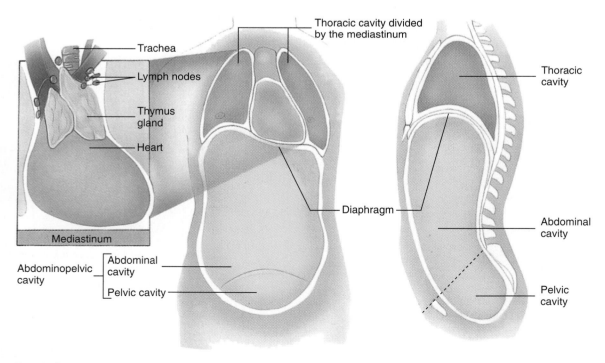

Trachea

Lymph nodes

Thymus gland

Heart

Mediastinum

Thoracic cavity divided by the mediastinum

Diaphragm

Abdominopelvic cavity

Abdominal cavity

Pelvic cavity

Thoracic cavity

Abdominal cavity

Pelvic cavity

Trunk Cavities
Figure 1.14

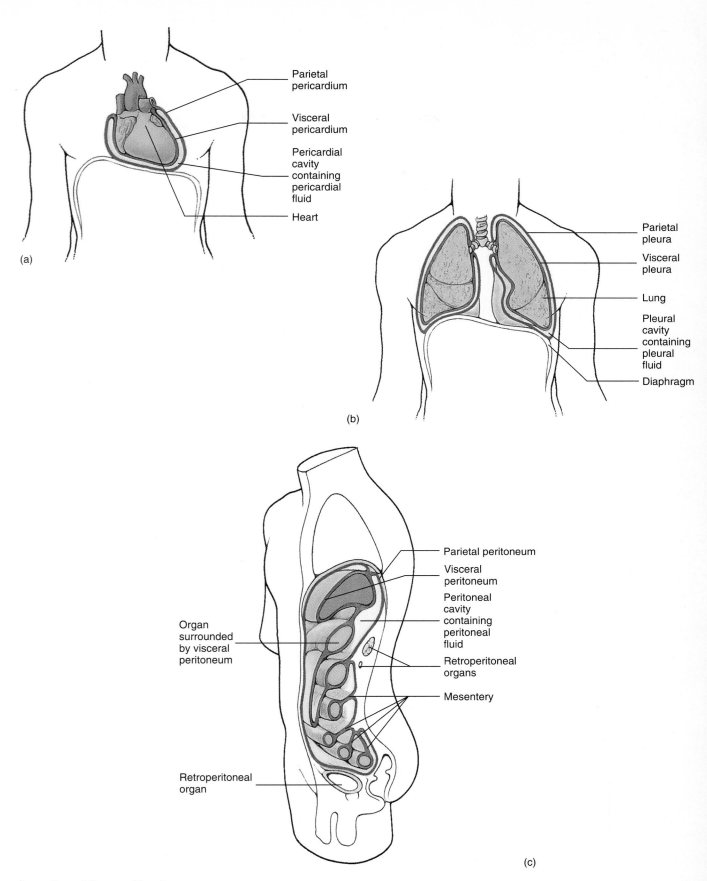

(a)

Parietal
pericardium

Visceral
pericardium

Pericardial
cavity
containing
pericardial
fluid

Heart

(b)

Parietal
pleura

Visceral
pleura

Lung

Pleural
cavity
containing
pleural
fluid

Diaphragm

Organ
surrounded
by visceral
peritoneum

Retroperitoneal
organ

Parietal peritoneum

Visceral
peritoneum

Peritoneal
cavity
containing
peritoneal
fluid

Retroperitoneal
organs

Mesentery

(c)

Location of Serous Membranes
Figure 1.16

7

Atom

Region occupied by negatively
charged electrons

Nucleus

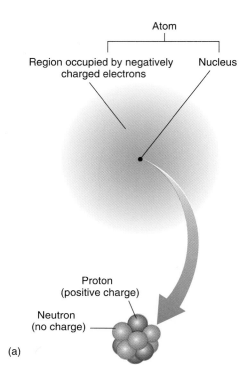

Proton
(positive charge)

Neutron
(no charge)

(a)

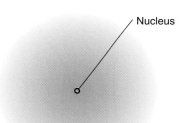

Nucleus

(b)

Features of an Atom
Figure 2.1

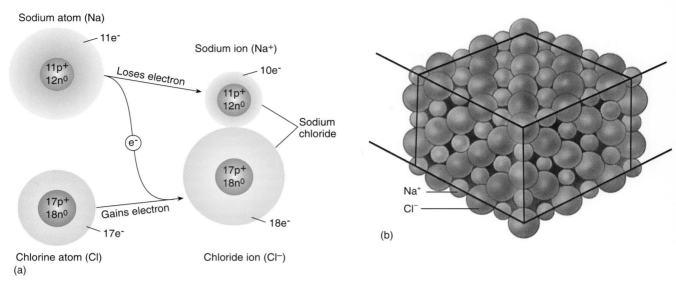

Sodium atom (Na)

11e⁻

11p⁺
12n⁰

Loses electron

Sodium ion (Na⁺)

10e⁻

11p⁺
12n⁰

Sodium
chloride

e⁻

17p⁺
18n⁰

Gains electron

17e⁻

Chlorine atom (Cl)

17p⁺
18n⁰

18e⁻

Chloride ion (Cl⁻)

(a)

Na⁺

Cl⁻

(b)

Ionic Bonds
Figure 2.4a, b

e⁻   e⁻

p⁺   p⁺

No interaction between the two hydrogen atoms because they are too far apart.

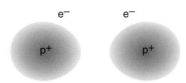

e⁻   e⁻

p⁺   p⁺

The positively charged nucleus of each hydrogen atom begins to attract the electron of the other.

e⁻

p⁺   •   p⁺

e⁻

A covalent bond is formed when the electrons are shared between the nuclei because the electrons are equally attracted to each nucleus.

Covalent Bonds
Figure 2.5

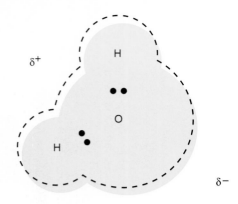

Polar Covalent Bonds
Figure 2.6

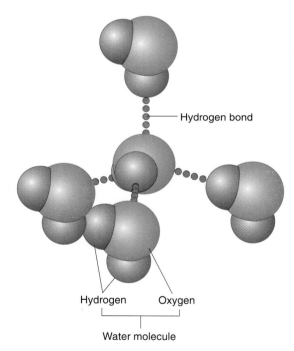

Hydrogen bond

Hydrogen   Oxygen

Water molecule

Hydrogen Bonds
Figure 2.7

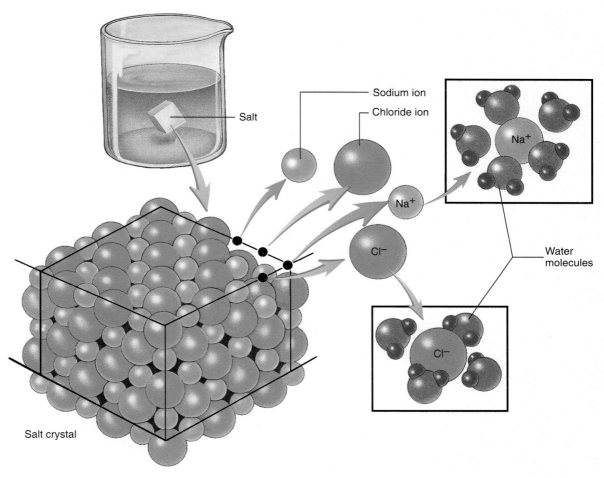

Salt

Sodium ion

Chloride ion

Na+

Cl−

Na+

Cl−

Water
molecules

Salt crystal

Dissociation
Figure 2.8

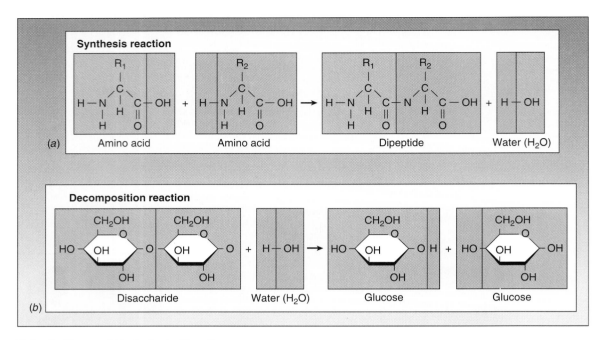

Dehydration and Hydrolysis Reactions
Figure 2.10

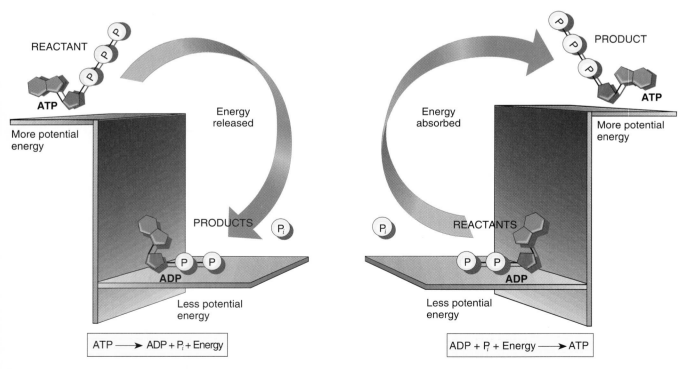

Energy and Chemical Reactions
Figure 2.11

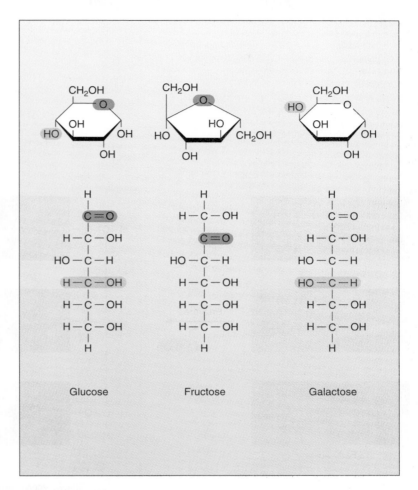

Monosaccharides
Figure 2.13

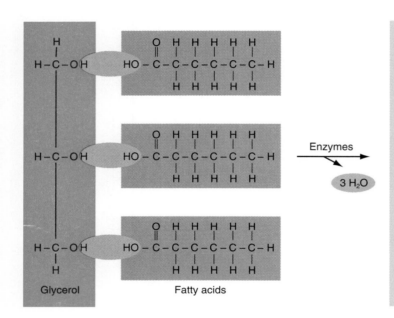

Glycerol   Fatty acids

Enzymes

3 H₂O

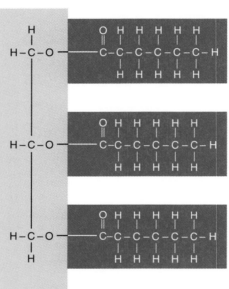

Triacylglycerol molecule

Triacylglycerols
Figure 2.15a

Primary structure—the amino acid sequence.

Amino acids

Peptide bond

Secondary structure with folding as a result of hydrogen bonding (dotted red lines).

Pleated sheet

Alpha helix

Tertiary structure with secondary folding caused by interactions within the polypeptide and its immediate environment.

Quaternary structure refers to the relationships between individual subunits.

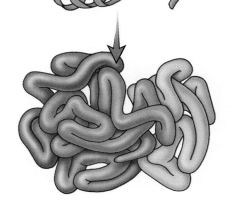

Protein Structure
Figure 2.21

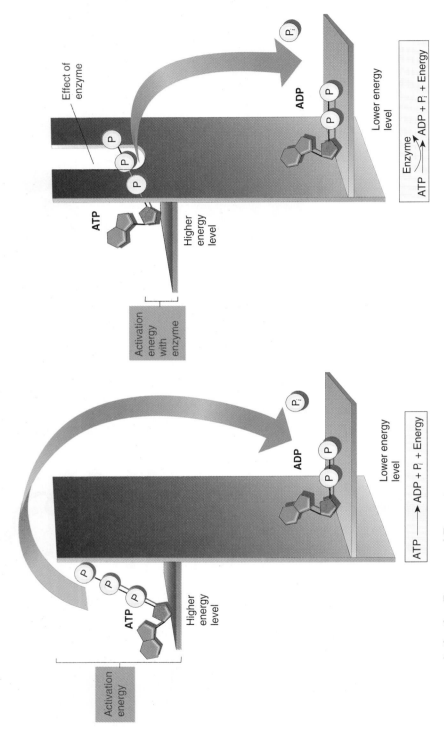

Activation Energy and Enzymes
Figure 2.22

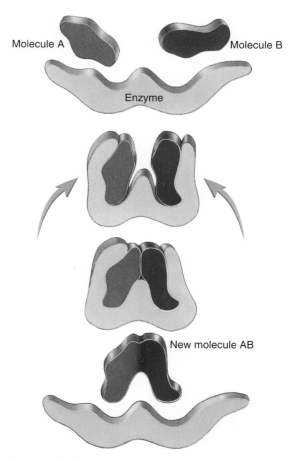

Molecule A

Molecule B

Enzyme

New molecule AB

Enzyme Action
Figure 2.23

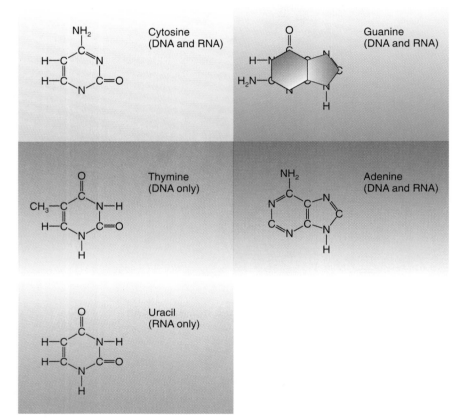

Pyrimidines

Cytosine
(DNA and RNA)

Thymine
(DNA only)

Uracil
(RNA only)

Purines

Guanine
(DNA and RNA)

Adenine
(DNA and RNA)

Nitrogenous Organic Bases
Figure 2.25

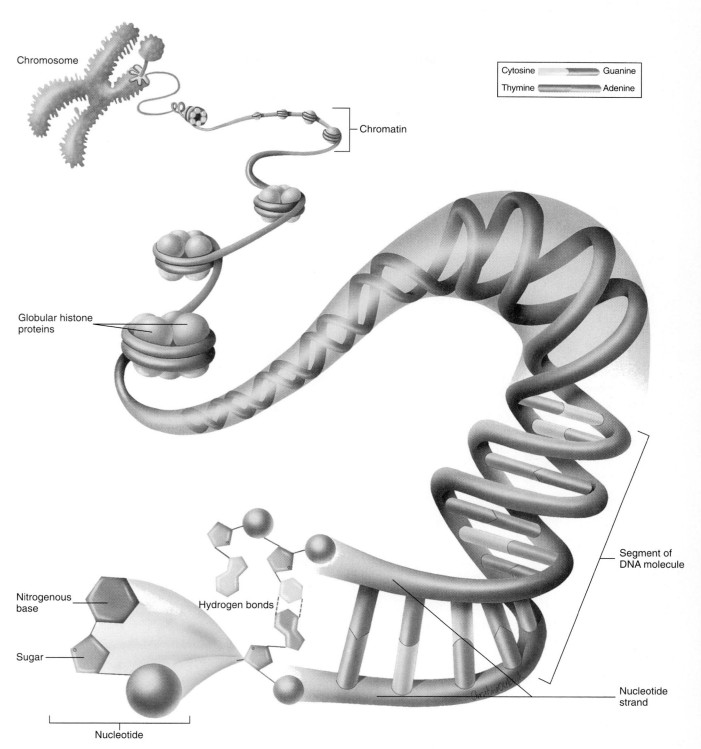

Chromosome

Cytosine ▭ Guanine
Thymine ▭ Adenine

Chromatin

Globular histone proteins

Segment of DNA molecule

Nitrogenous base

Hydrogen bonds

Sugar

Nucleotide strand

Nucleotide

Structure of DNA
Figure 2.26

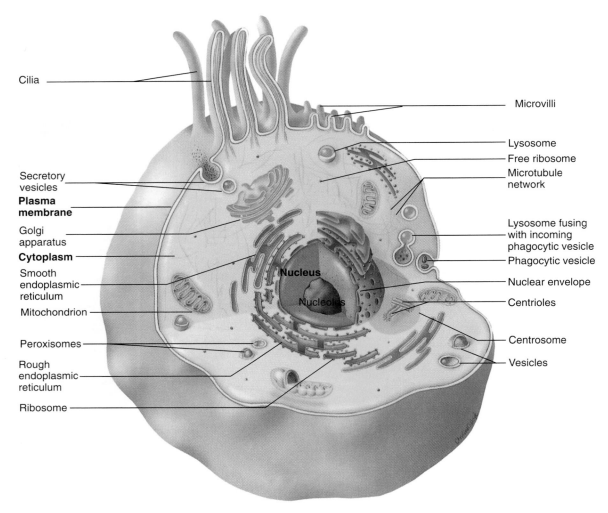

Cilia

Microvilli

Lysosome

Free ribosome

Microtubule network

Secretory vesicles

Plasma membrane

Golgi apparatus

Cytoplasm

Smooth endoplasmic reticulum

Mitochondrion

Peroxisomes

Rough endoplasmic reticulum

Ribosome

Lysosome fusing with incoming phagocytic vesicle

Phagocytic vesicle

Nuclear envelope

Centrioles

Centrosome

Vesicles

Nucleus

Nucleolus

The Cell
Figure 3.1

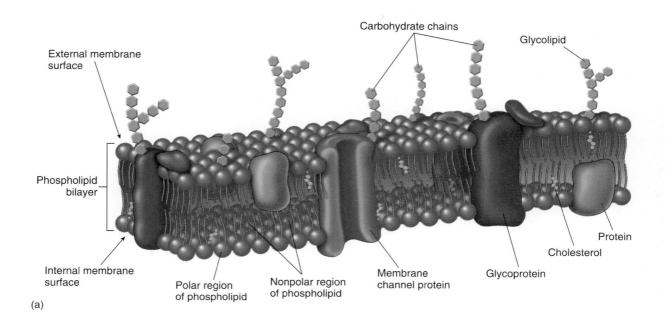

External membrane surface

Carbohydrate chains

Glycolipid

Phospholipid bilayer

Internal membrane surface

Polar region of phospholipid

Nonpolar region of phospholipid

Membrane channel protein

Glycoprotein

Cholesterol

Protein

(a)

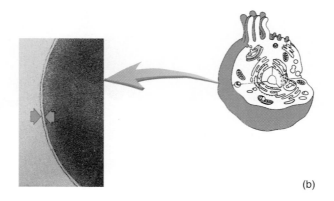

(b)

Cell Membrane
Figure 3.2

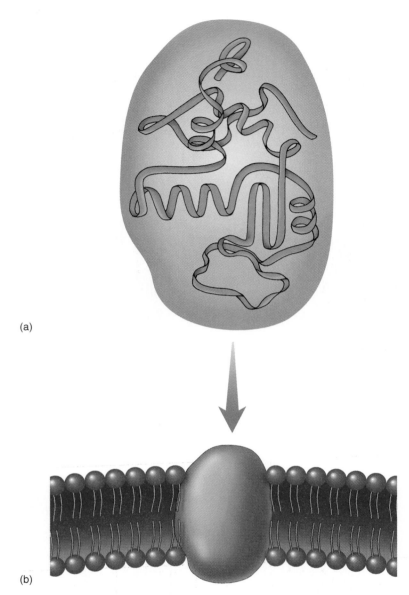

(a)

(b)

Globular Proteins in the Cell Membrane
Figure 3.3

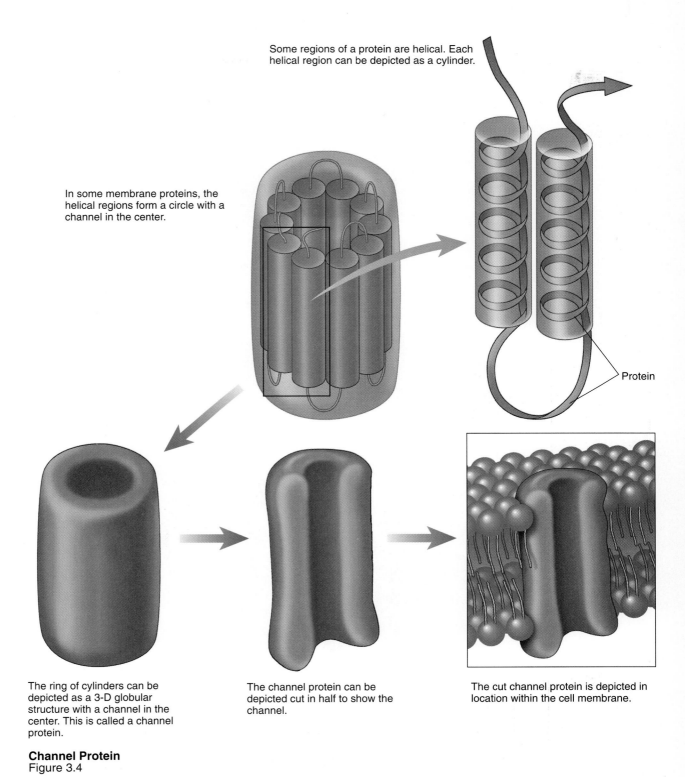

Some regions of a protein are helical. Each helical region can be depicted as a cylinder.

In some membrane proteins, the helical regions form a circle with a channel in the center.

Protein

The ring of cylinders can be depicted as a 3-D globular structure with a channel in the center. This is called a channel protein.

The channel protein can be depicted cut in half to show the channel.

The cut channel protein is depicted in location within the cell membrane.

Channel Protein
Figure 3.4

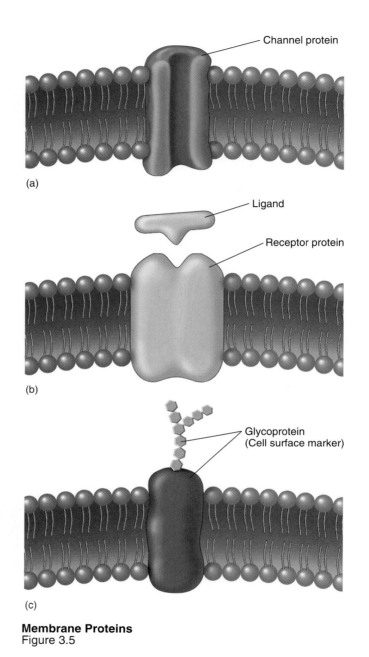

(a)

(b)

(c)

Membrane Proteins
Figure 3.5

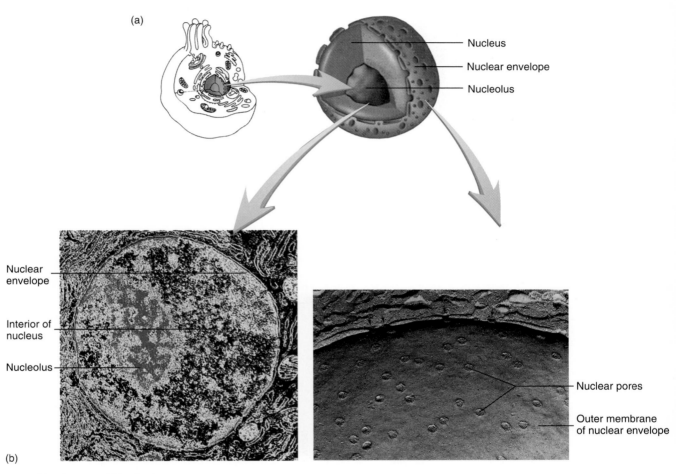

(a)

Nucleus

Nuclear envelope

Nucleolus

Nuclear envelope

Interior of nucleus

Nucleolus

(b)

Nuclear pores

Outer membrane of nuclear envelope

Nucleus and Nucleolus
Figure 3.6

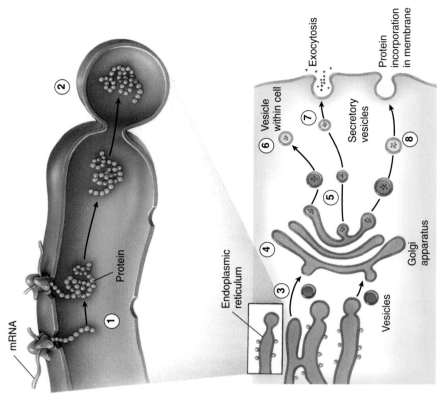

1. Proteins are produced at ribosomes on the surface of the rough endoplasmic reticulum and are transferred into the cisterna as they are produced.

2. The proteins are surrounded by vesicles that form from the membrane of the endoplasmic reticulum.

3. The vesicle moves from the endoplasmic reticulum to the Golgi apparatus, fuses with the membrane of the Golgi apparatus, and releases the protein into the cisterna of the Golgi apparatus.

4. The Golgi apparatus concentrates and, in some cases, modifies the proteins into glycoproteins or lipoproteins.

5. The proteins are packaged into vesicles that form from the membrane of the Golgi apparatus.

6. Some vesicles contain enzymes that are used within the cell.

7. Some vesicles carry proteins to the plasma membrane where the proteins are secreted from the cell by exocytosis.

8. Some vesicles contain proteins that become part of the plasma membrane.

Function of the Golgi Apparatus
Figure 3.11

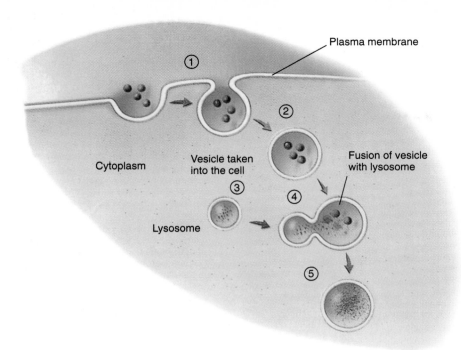

Plasma membrane

Cytoplasm

Vesicle taken
into the cell

Lysosome

Fusion of vesicle
with lysosome

① ② ③ ④ ⑤

1. Vesicles containing materials
   from outside the cell are
   taken into the cell.

2. The vesicle is pinched off from
   the plasma membrane and
   becomes a separate vesicle.

3. A lysosome approaches the vesicle.
4. The lysosome fuses with the vesicle.

5. The enzymes from the lysosome mix with
   the material in the vesicle, and the
   enzymes digest the material.

Action of Lysosomes
Figure 3.12

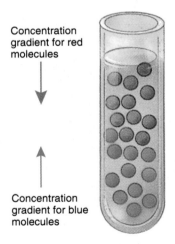

Concentration
gradient for red
molecules

Concentration
gradient for blue
molecules

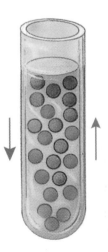

One solution (red balls representing
one type of molecule) is layered onto
a second solution (blue balls
representing a second type of
molecule). There is a concentration
gradient for the red molecules from
the red solution into the blue solution
because there are no red molecules
in the blue solution. There is also a
concentration gradient for the blue
molecules from the blue solution into
the red solution because there are
no blue molecules in the red
solution.

Red molecules move down their
concentration gradient into the
blue solution (red arrow), and
the blue molecules move down
their concentration gradient into
the red solution (blue arrow).

Red and blue molecules are
distributed evenly throughout the
solution. Even though the red and
blue molecules continue to move
randomly, an equilibrium exists,
and no net movement occurs
because no concentration
gradient exists.

Diffusion
Figure 3.17

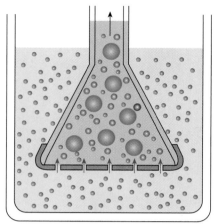

Because the tube contains salt (green and red spheres) as well as water molecules (blue spheres), the tube has proportionately less water than is in the beaker, which contains only water. The water molecules diffuse with their concentration gradient into the tube (blue arrows). Because the salt molecules cannot leave the tube, the total fluid volume inside the tube increases, and fluid moves up the glass tube (black arrow) as a result of osmosis.

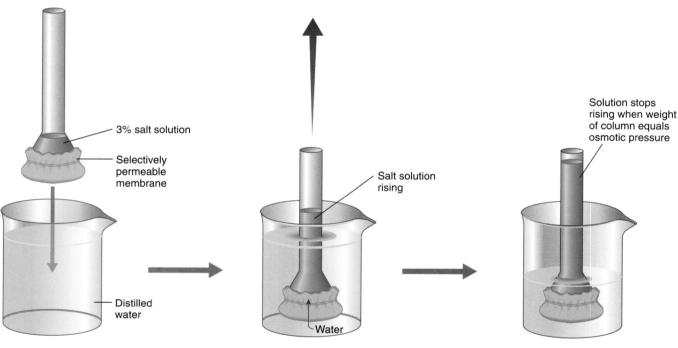

3% salt solution

Selectively permeable membrane

Distilled water

Salt solution rising

Water

Solution stops rising when weight of column equals osmotic pressure

The end of a tube containing a 3% salt solution (green) is closed at one end with a selectively permeable membrane, which allows water molecules to pass through it but retains the salt molecules within the tube.

The tube is immersed in distilled water. Water moves into the tube by osmosis (see inset above).

Water continues to move into the tube until the weight of the column of water in the tube (hydrostatic pressure) exerts a downward force equal to the osmotic force moving water molecules into the tube. The hydrostatic pressure that prevents net movement of water into the tube is termed the osmotic pressure of the solution in the tube.

Osmosis
Figure 3.18

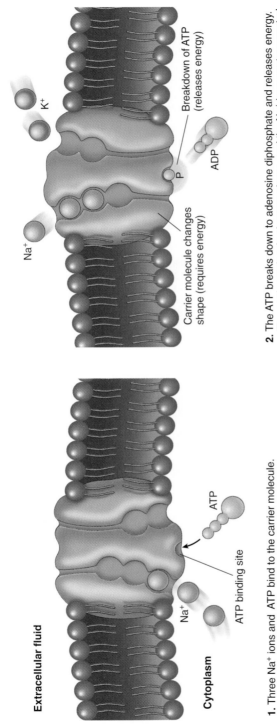

Extracellular fluid

Cytoplasm

K⁺

Na⁺

Breakdown of ATP (releases energy)

P

ADP

Carrier molecule changes shape (requires energy)

2. The ATP breaks down to adenosine diphosphate and releases energy. The carrier molecule changes shape, and the Na⁺ ions are transported across the membrane.

Na⁺

ATP

ATP binding site

1. Three Na⁺ ions and ATP bind to the carrier molecule.

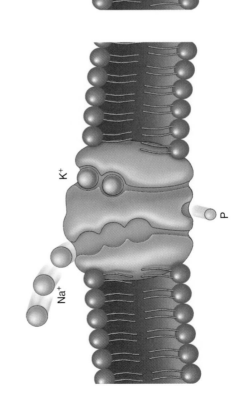

Carrier molecule resumes original shape

K⁺

4. The carrier molecule changes shape, transporting K⁺ ions across the membrane, and the K⁺ ions diffuse away from the carrier molecule. The carrier molecule can again bind to Na⁺ ions and ATP.

K⁺

Na⁺

P

3. The Na⁺ ions diffuse away from the carrier molecule, two K⁺ ions bind to the carrier molecule, and the phosphate is released.

Sodium-Potassium Exchange Pump
Figure 3.23

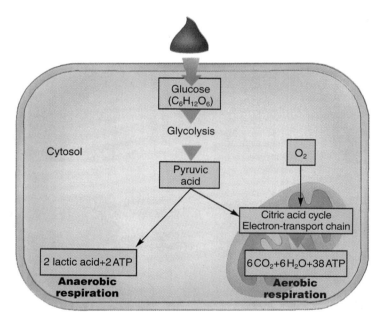

Overview of Cell Metabolism
Figure 3.27

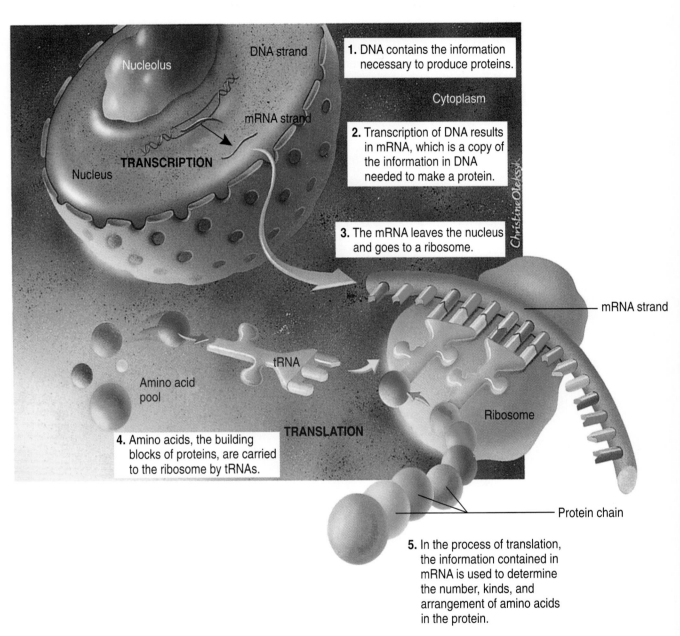

Inside the figure:

Nucleolus

DNA strand

mRNA strand

1. DNA contains the information necessary to produce proteins.

Cytoplasm

2. Transcription of DNA results in mRNA, which is a copy of the information in DNA needed to make a protein.

TRANSCRIPTION

Nucleus

3. The mRNA leaves the nucleus and goes to a ribosome.

ChristineOleksyk

mRNA strand

tRNA

Amino acid pool

Ribosome

TRANSLATION

4. Amino acids, the building blocks of proteins, are carried to the ribosome by tRNAs.

Protein chain

5. In the process of translation, the information contained in mRNA is used to determine the number, kinds, and arrangement of amino acids in the protein.

Overview of Protein Synthesis
Figure 3.28

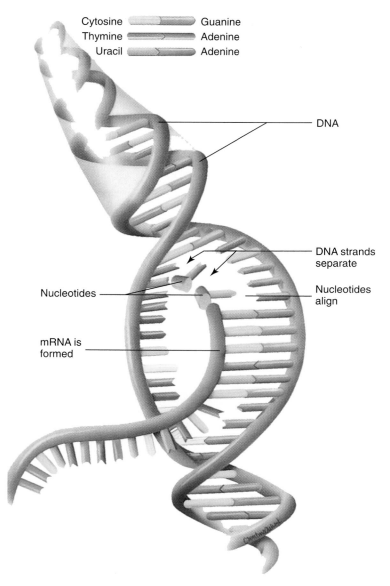

Formation of mRNA by Transcription of DNA
Figure 3.29

1. To start protein synthesis a ribosome binds to mRNA. The ribosome also has two binding sites for tRNA, one of which is occupied by a tRNA with its amino acid. Note that the codon of mRNA and the anticodon of tRNA are aligned and joined. The other tRNA binding site is open.

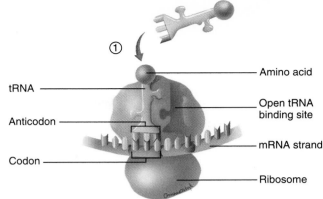

tRNA

Anticodon

Codon

Amino acid

Open tRNA binding site

mRNA strand

Ribosome

2. By occupying the open tRNA binding site the next tRNA is properly aligned with mRNA and with the other tRNA.

3. An enzyme within the ribosome catalyzes a synthesis reaction to form a peptide bond between the amino acids. Note that the amino acids are now associated with only one of the tRNAs.

4. The ribosome shifts position by three nucleotides. The tRNA without the amino acid is released from the ribosome, and the tRNA with the amino acids takes its position. A tRNA binding site is left open by the shift. Additional amino acids can be added by repeating steps 2 to 4. Eventually a stop codon in the mRNA ends the production of the protein, which is released from the ribosome.

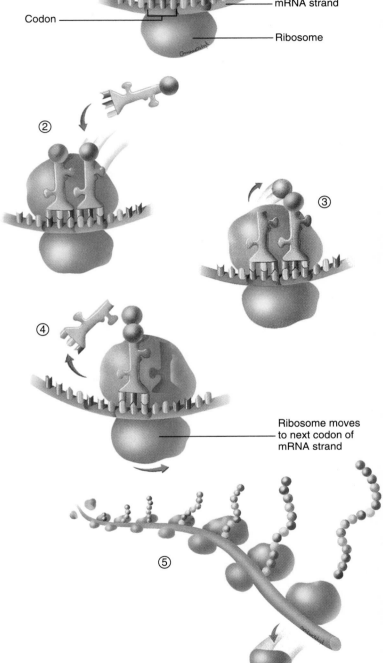

Ribosome moves to next codon of mRNA strand

5. This is an overview of protein synthesis.

Translation of mRNA to Produce a Protein
Figure 3.31

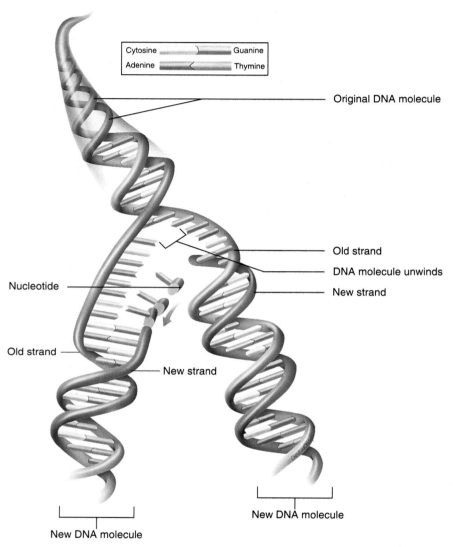

| Cytosine | Guanine |
| Adenine | Thymine |

Original DNA molecule

Old strand

DNA molecule unwinds

Nucleotide

New strand

Old strand

New strand

New DNA molecule

New DNA molecule

Replication of DNA
Figure 3.33

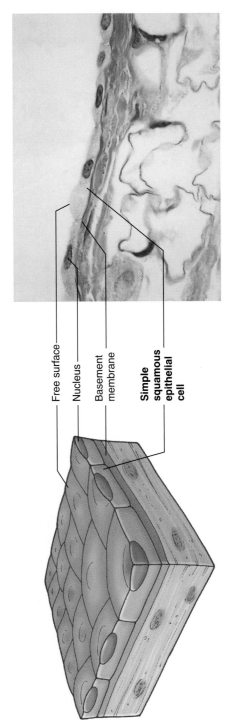

Types of Epithelium

Simple squamous epithelium

Location: Lining of blood and lymph vessels (endothelium) and small ducts, alveoli of the lungs, loop of Henle in kidney tubules, lining of serous membranes (mesothelium), and inner surface of the eardrum.

Structure: Single layer of flat, often hexagonal cells. The nuclei appear as bumps when viewed as a cross section because the cells are so flat.

Function: Diffusion, filtration, some protection against friction, secretion, and absorption.

(a)

Simple Squamous Epithelium, Stratified Squamous Epithelium
Figure 4.2a, d, *(Continued)*

Free surface

Nucleus

Basement membrane

Simple squamous epithelial cell

Capillary epithelial cells

Nucleus

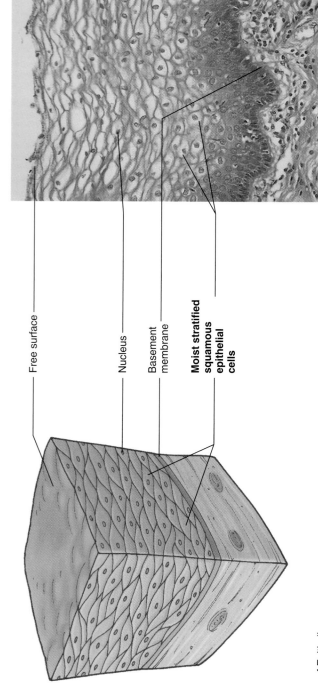

Types of Epithelium

Stratified squamous epithelium

Location: Moist—mouth, throat, larynx, esophagus, anus, vagina, inferior urethra, and cornea. Keratinized—skin.

Structure: Multiple layers of cells that are cuboidal in the basal layer and progressively flattened toward the surface. The epithelium can be moist or keratinized. In moist stratified squamous epithelium the surface cells retain a nucleus and cytoplasm. In keratinized cells the cytoplasm is replaced by keratin, and the cells are dead.
(d)

Function: Protection against abrasion and infection.

Figure 4.2a, d

Free surface

Nucleus

Basement membrane

Moist stratified squamous epithelial cells

Cornea

Mouth

Esophagus

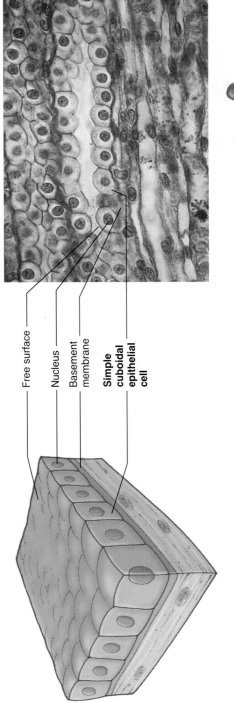

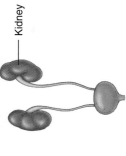

Kidney

Free surface

Nucleus

Basement membrane

Simple cuboidal epithelial cell

Types of Epithelium

Simple cuboidal epithelium

Location: Glands and their ducts, terminal bronchioles of lungs, kidney tubules, choroid plexus of the brain, and surface of the ovaries.

Structure: Single layer of cube-shaped cells. Some cells have cilia (terminal bronchioles) or microvilli (kidney tubules).

Function: Movement of mucus-containing particles out of the terminal bronchioles by ciliated cells. Absorption and secretion by cells of the kidney tubules. Secretion by cells of the choroid plexus and glands.
(b)

Simple Cuboidal Epithelium, Stratified Cuboidal Epithelium
Figure 4.2b, e, *(Continued)*

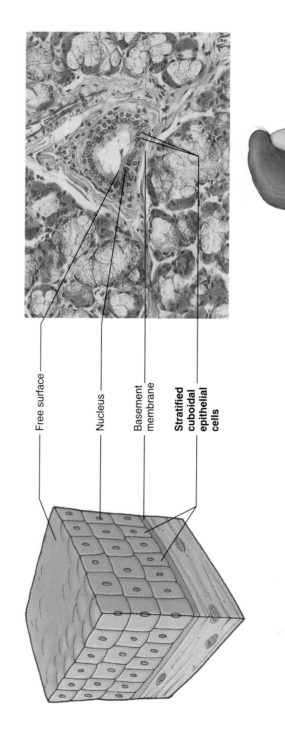

Types of Epithelium

Stratified cuboidal epithelium

Location: Sweat gland ducts, ovarian follicular cells, and salivary gland ducts.

Structure: Multiple layers of somewhat cube-shaped cells.

Function: Secretion, absorption, and protection against infection.
(e)

Figure 4.2b, e

Free surface

Nucleus

Basement membrane

Stratified cuboidal epithelial cells

Parotid gland duct

Sublingual gland duct

Submandibular gland duct

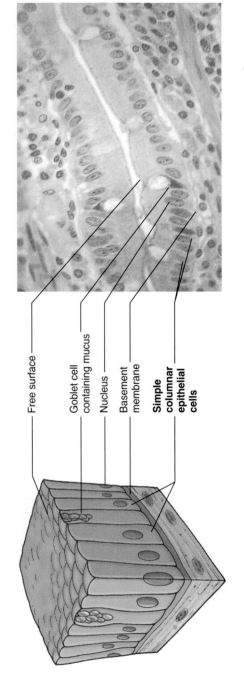

Types of Epithelium

Simple columnar epithelium

Location: Glands and some ducts, bronchioles of lungs, auditory tubes, uterus, uterine tubes, stomach, intestines, gallbladder, bile ducts, and ventricles of the brain.

Structure: Single layer of tall, narrow cells. Some cells have cilia (bronchioles of lungs, auditory tubes, uterine tubes, and uterus) or microvilli (intestines).

Function: Movement of particles out of the bronchioles of the lungs; partially responsible for the movement of the oocyte through the uterine tubes by ciliated cells. Secretion by cells of the glands, the stomach, and the intestine. Absorption by cells of the intestine.

(c)

Simple Columnar Epithelium, Stratified Columnar Epithelium
Figure 4.2c, f *(Continued)*

Free surface

Goblet cell containing mucus

Nucleus

Basement membrane

Simple columnar epithelial cells

Lining of stomach and intestines

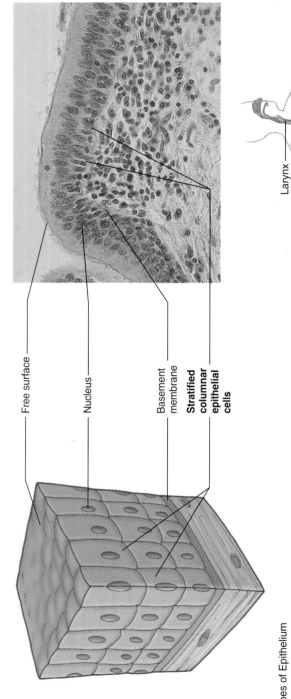

Types of Epithelium

Stratified columnar epithelium

Location: Mammary gland duct, larynx, and a portion of the male urethra.

Structure: Multiple layers of cells, with tall, thin cells resting on layers of more cuboidal cells. The cells are ciliated in the larynx.

Function: Protection and secretion.

(f)

Simple Columnar Epithelium, Stratified Columnar Epithelium
Figure 4.2c, f *(Continued)*

Free surface

Nucleus

Basement membrane

Stratified columnar epithelial cells

Larynx

Types of Epithelium

Pseudostratified columnar epithelium

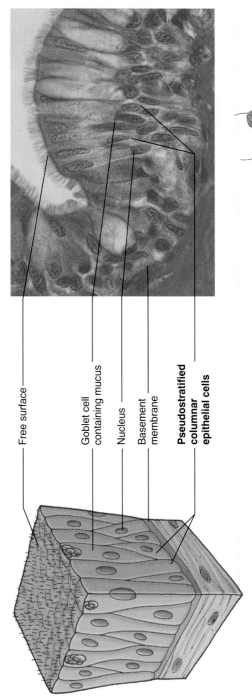

Location: Larynx, nasal cavity, paranasal sinuses, pharynx, auditory tubes, trachea, and bronchi of the lungs.

Structure: Single layer of cells. All the cells are attached to the basement membrane. Some cells are tall and thin and reach the free surface, and other do not. The nuclei of these cells are at different levels and appear stratified. The cells are almost always ciliated and are associated with goblet cells.

Function: Movement of fluid (often mucus) that contains foreign particles.
(g)

Pseudostratified Columnar Epithelium, Transitional Epithelium
Figure 4.2g, h (continued)

Free surface

Goblet cell containing mucus

Nucleus

Basement membrane

Pseudostratified columnar epithelial cells

Trachea

Bronchus

Types of Epithelium

Transitional epithelium

Location: Urinary bladder, ureters, and superior urethra.

Structure: Stratified cells that appear cubelike when the organ or tube is relaxed and appear squamouslike when the organ or tube is distended by fluid.

Function: Formation of a permeability barrier and protection against the caustic effect of urine. Accommodation of fluid content fluctuations in organ or tube.

(h)

Pseudostratified Columnar Epithelium, Transitional Epithelium
Figure 4.2g, h

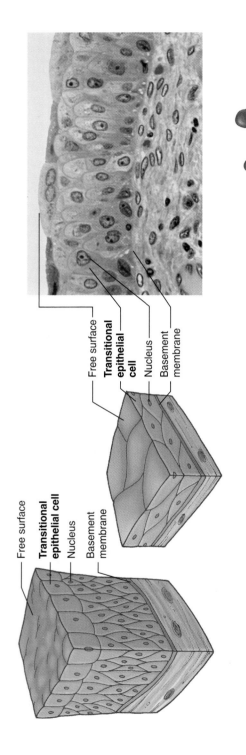

Free surface

Transitional epithelial cell

Nucleus

Basement membrane

Free surface

Transitional epithelial cell

Nucleus

Basement membrane

Ureter

Urinary bladder

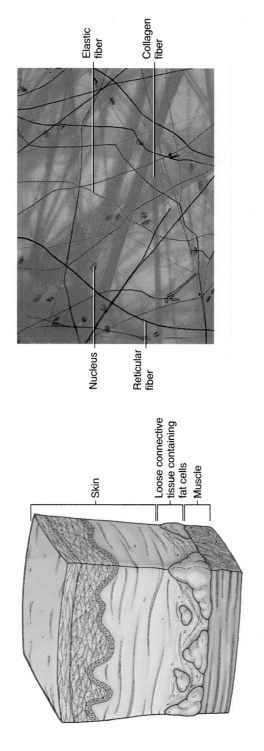

Elastic fiber

Collagen fiber

Nucleus

Reticular fiber

Connective Tissues

Areolar, or loose, connective tissue.

Location: Widely distributed throughout the body; substance on which epithelial basement membranes rest; packing between glands, muscles, and nerves. Attaches the skin to underlying tissues.

Structure: Cells (for example, fibroblasts, macrophages, and lymphocytes) within a fine network of mostly collagen fibers. Often merges with denser connective tissue.

Function: Loose packing, support, and nourishment for the structures with which it is associated.
(a)

Loose Connective Tissue, Reticular Tissue
Figure 4.6a, g (continued)

Skin

Loose connective tissue containing fat cells

Muscle

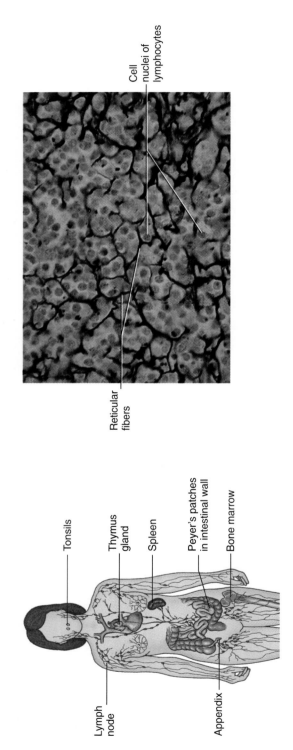

Cell
nuclei of
lymphocytes

Reticular
fibers

Tonsils

Thymus
gland

Spleen

Peyer's patches
in intestinal wall

Bone marrow

Lymph
node

Appendix

Connective Tissues

Reticular tissue.

Location: Within the lymph nodes, spleen, and bone marrow.

Structure: Fine network of reticular fibers irregularly arranged.

Function: Provides a superstructure for the lymphatic and hemopoietic tissues.
(g)

Loose Connective Tissue, Reticular Tissue
Figure 4.6a, g

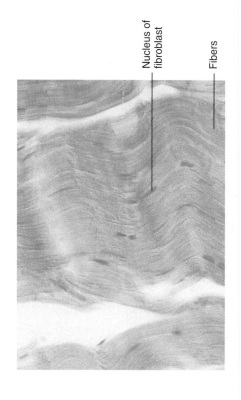

Nucleus of fibroblast

Fibers

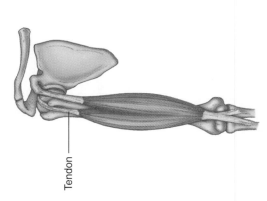

Tendon

Connective Tissues

Dense regular collagenous connective tissue.

Location: Tendons (attach muscle to bone) and ligaments (attach bones to each other).

Structure: Matrix composed of collagen fibers running in somewhat the same direction.

Function: Ability to withstand great pulling forces exerted in the direction of fiber orientation, great tensile strength, and stretch resistance.
(b)

Dense Regular Collagenous Connective Tissue, Dense Regular Elastic Connective Tissue
Figure 4.6b, c (continued)

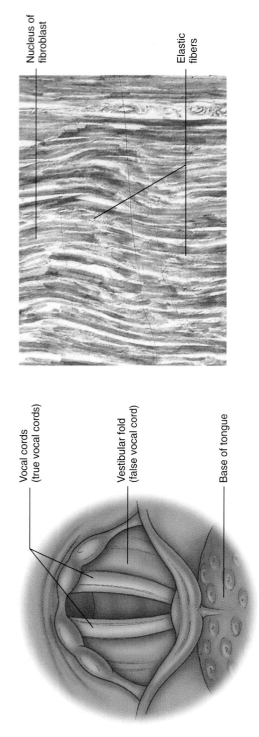

Nucleus of
fibroblast

Elastic
fibers

Vocal cords
(true vocal cords)

Vestibular fold
(false vocal cord)

Base of tongue

Connective Tissues

Dense regular elastic connective tissue.

Location: Ligaments between the vertebrae and along the dorsal aspect of the neck (nucha) and in the vocal cords.

Structure: Matrix composed of regularly arranged collagen fibers and elastin fibers.

Function: Capable of stretching and recoiling like a rubber band with strength in the direction of fiber orientation.

(c)

Dense Regular Collagenous Connective Tissue, Dense Regular Elastic Connective Tissue
Figure 4.6b, c

Connective Tissues

Dense irregular collagenous connective tissue.

Location: Sheaths; dermis of the skin; organ capsules and septa; outer covering of body tubes.

Structure: Matrix composed of collagen fibers that run in all directions or in alternating planes of fibers oriented in a somewhat single direction.

Function: Tensile strength capable of withstanding stretching in all directions.

(d)

Dense Irregular Collagenous Connective Tissue, Dense Irregular Elastic Connective Tissue
Figure 4.6d, e (continued)

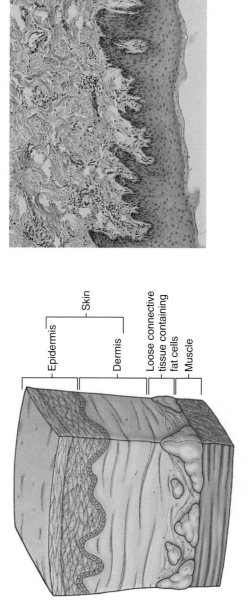

Dense irregular collagenous connective tissue

Epidermis
Skin
Dermis
Loose connective tissue containing fat cells
Muscle

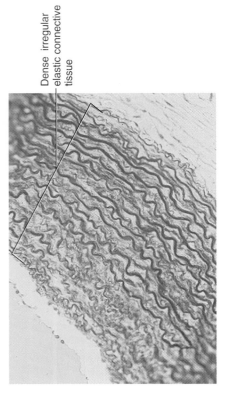

Dense irregular
elastic connective
tissue

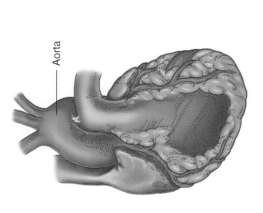

Aorta

Connective Tissues

Dense irregular elastic connective tissue.

Location: Elastic arteries.

Structure: Matrix composed of bundles and sheets of collagenous and elastin fibers oriented in multiple directions.

Function: Capable of strength with stretching and recoil in several directions.

(e)

Dense Irregular Collagenous Connective Tissue, Dense Irregular Elastic Connective Tissue

Figure 4.6d, e

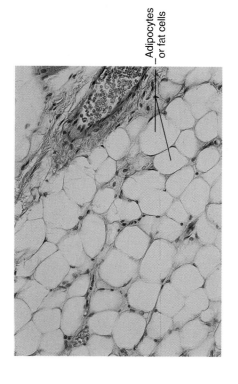

Adipocytes
or fat cells

Connective Tissues

Adipose tissue.

Location: Predominantly in subcutaneous areas, mesenteries, renal, pelvis, around kidneys, attached to the surface of the colon, mammary glands, and in loose connective tissue that penetrates into spaces and crevices.

Structure: Little extracellular material surrounding cells. The adipocytes, or fat cells, are so full of lipid that the cytoplasm is pushed to the periphery of the cell.

Function: Packing material, thermal insulator, energy storage, and protection of organs against injury from being bumped or jarred.

(f)

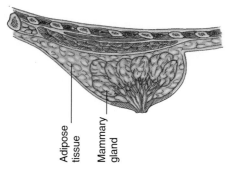

Adipose
tissue

Mammary
gland

Adipose Tissue, Hyaline Cartilage
Figure 4.6f, i (continued)

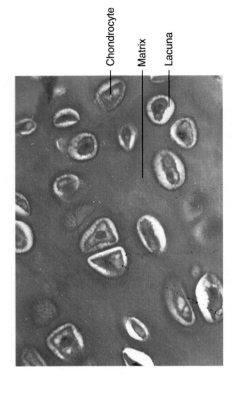

Chondrocyte

Matrix

Lacuna

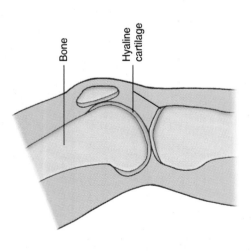

Bone

Hyaline
cartilage

Connective Tissues

Hyaline cartilage.

Location: Growing long bones, cartilage rings of the respiratory system, costal cartilage of ribs, nasal cartilage, articulating surface of bones, and the embryonic skeleton.

Structure: Collagen fibers are small and evenly dispersed in the matrix, making the matrix appear transparent. The cartilage cells, or chondrocytes, are formed in spaces, or lacunae, within the rigid matrix.

Function: Allows growth of long bones. Provides rigidity with some flexibility in the trachea, bronchi ribs, and nose. Forms rugged, smooth, yet somewhat flexible articulating surfaces. forms the embryonic skeleton. (i)

Adipose Tissue, Hyaline Cartilage
Figure 4.6f, i

Connective Tissues

Fibrocartilage.

Location: Intervertebral disks, symphysis pubis, articular disks (for example, knee and temporomandibular [jaw] joints).

Structure: Collagenous fibers similar to those in hyaline cartilage. The fibers are more numerous than in other cartilages and are arranged in thick bundles.

Function: Somewhat flexible and capable of withstanding considerable pressure. Connects structures subjected to great pressure.
(j)

Fibrocartilage, Elastic Cartilage
Figure 4.6j, k, *(Continued)*

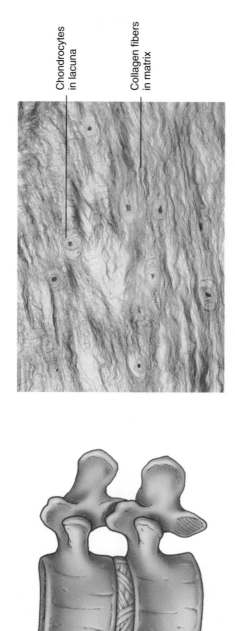

Chondrocytes
in lacuna

Collagen fibers
in matrix

Intervertebral
disk

Elastic fibers
in matrix

Chondrocyte
in lacuna

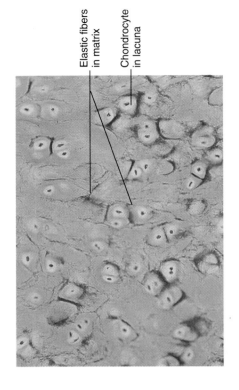

Connective Tissues

Elastic cartilage.

Location: External ear, epiglottis, and auditory tubes.

Structure: Similar to hyaline cartilage, but matrix also contains elastin fibers.

Function: Provides rigidity with even more flexibility than hyaline cartilage because elastic fibers return to their original shape after being stretched.
(k)

Figure 4.6j, k

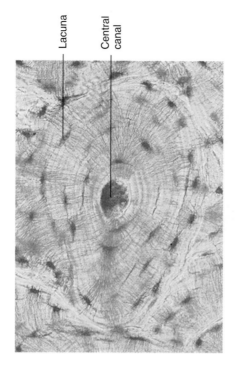

Lacuna

Central canal

Connective Tissues

Compact bone.

Location: Outer portions of all bones and the shafts of long bones.

Structure: Hard, bony matrix predominates. Many osteocytes are located within lacunae that are distributed in a circular fashion around the central canals. Small passageways connect adjacent lacunae.

Function: Provides great strength and support. Forms a solid outer shell on bones that keeps them from being easily broken or punctured.

Compact Bone
Figure 4.6 m

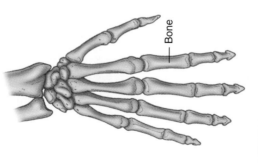

Bone

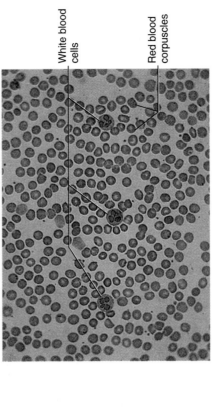

White blood cells

Red blood corpuscles

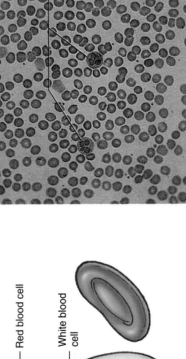

Red blood cell

White blood cell

Connective Tissues

Blood.

Location: Within the blood vessels. Produced by the hemopoietic tissues. White blood cells frequently leave the blood vessels and enter the interstitial spaces.

Structure: Blood cells and a fluid matrix.

Function: Transports oxygen, carbon dioxide, hormones, nutrients, waste products, and other substances. Protects the body from infections and is involved in temperature regulation.

Blood
Figure 4.6 n

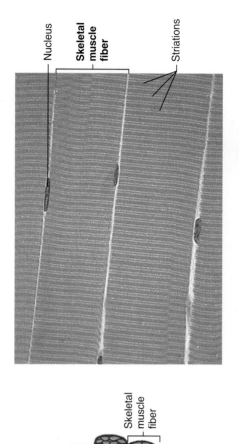

Nucleus

Skeletal muscle fiber

Striations

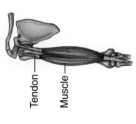

Tendon

Muscle

Nucleus

Skeletal muscle fiber

Striations

Muscle Tissue

Skeletal muscle.

Location: Attached to bone.

Structure: Appears striated. Cells are large, long, and cylindrical with several peripherally located nuclei in each cell.

Function: Movement of the body; under voluntary control.

Skeletal Muscle
Figure 4.7a

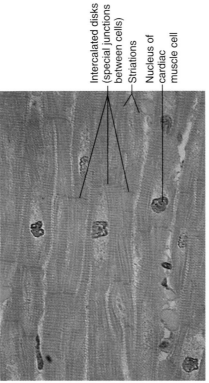

Intercalated disks (special junctions between cells)

Striations

Nucleus of cardiac muscle cell

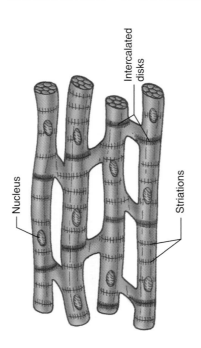

Nucleus

Intercalated disks

Striations

Muscle Tissue

Cardiac muscle.

Location: Heart.

Structure: Appears striated. Cells are cylindrical and branching with a single, centrally located nucleus. Cells are connected to each other by specialized gap junctions called intercalated disks.

Function: Pumps the blood; under involuntary control.

Cardiac Muscle
Figure 4.7b

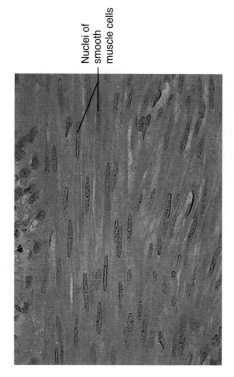

Nuclei of smooth muscle cells

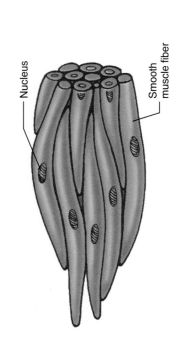

Nucleus

Smooth muscle fiber

Wall of stomach

Wall of colon

Wall of small intestine

Muscle Tissue

Smooth muscle.

Location: In the walls of hollow organs, iris of the eye, skin (attached to hair), and glands.

Structure: No striations. Cells are spindle-shaped with a single, centrally located nucleus.

Function: Regulates the size of organs, forces fluid through tubes, controls amount of light entering the eye, and produces "goose flesh" in the skin; under involuntary control.

Smooth Muscle
Figure 4.7c

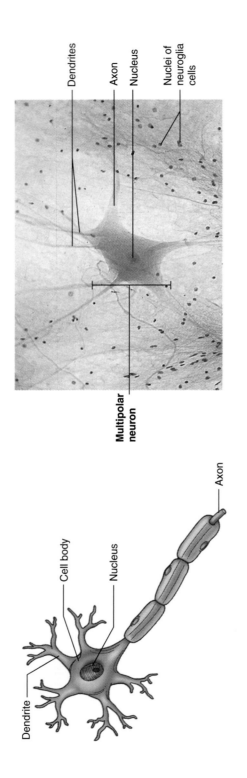
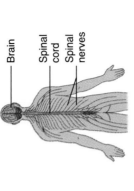

Multipolar neuron

Neurons

Multipolar neuron.

Location: Cell bodies—in the brain, spinal cord, or ganglia; cell processes—all parts of the body.

Structure: Mainly relatively large cells with a variety of shapes; characterized by many cell processes.

Function: Conduct action potentials, store "information," and in some way integrate and evaluate data.

Multipolar Neurons
Figure 4.8a

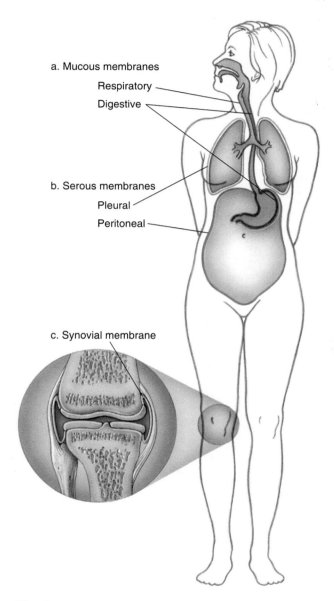

a. Mucous membranes

Respiratory

Digestive

b. Serous membranes

Pleural

Peritoneal

c. Synovial membrane

Membranes
Figure 4.10

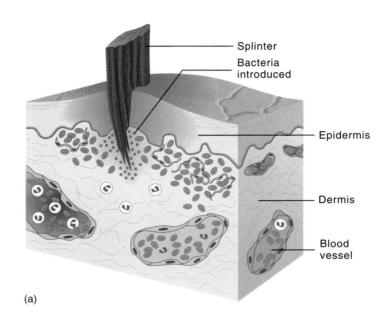

Splinter

Bacteria introduced

Epidermis

Dermis

Blood vessel

(a)

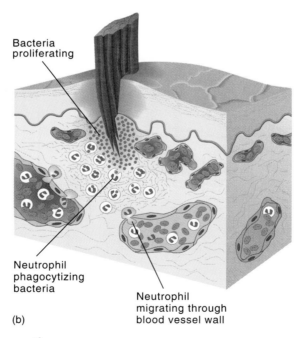

Bacteria proliferating

Neutrophil phagocytizing bacteria

Neutrophil migrating through blood vessel wall

(b)

Inflammation
Figure 4.11

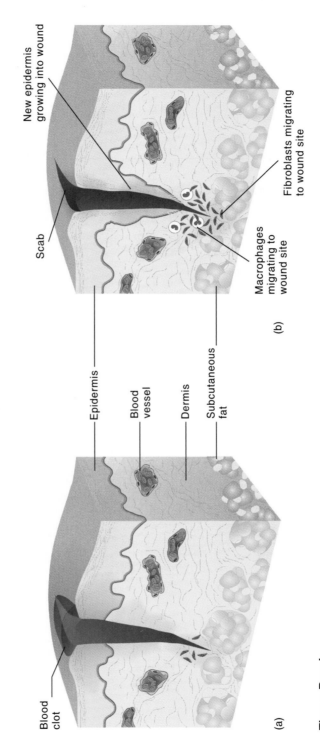

New epidermis growing into wound

Scab

Fibroblasts migrating to wound site

Macrophages migrating to wound site

(b)

Epidermis

Blood vessel

Dermis

Subcutaneous fat

Blood clot

(a)

Tissue Repair
Figure 4.12a, b

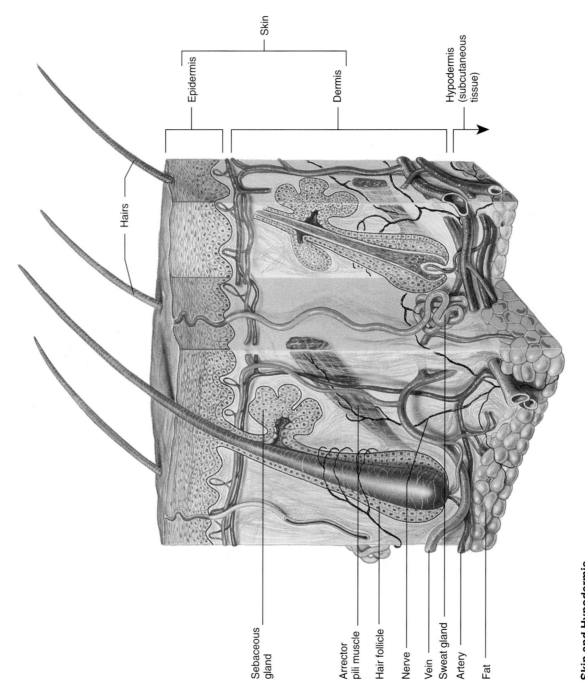

Skin

Epidermis

Dermis

Hypodermis
(subcutaneous
tissue)

Hairs

Sebaceous
gland

Arrector
pili muscle

Hair follicle

Nerve

Vein

Sweat gland

Artery

Fat

Skin and Hypodermis
Figure 5.1

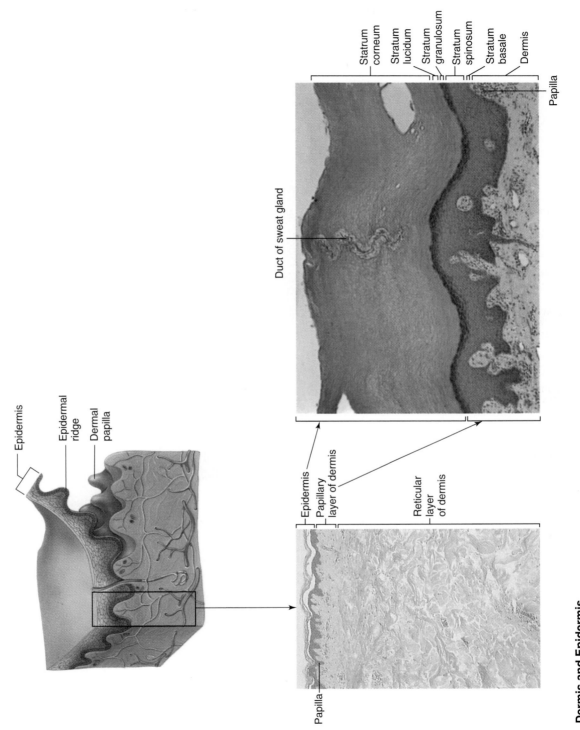

Statrum corneum
Stratum lucidum
Stratum granulosum
Stratum spinosum
Stratum basale
Dermis

Papilla

Duct of sweat gland

Epidermis
Epidermal ridge
Dermal papilla

Epidermis
Papillary layer of dermis

Reticular layer of dermis

Papilla

Dermis and Epidermis
Figure 5.2

Superficial

5. Stratum corneum
  Dead cells with a hard protein envelope;
  the cells contain keratin and are surrounded
  by lipids.

4. Stratum lucidum
  Dead cells lie within dispersed keratohyalin.

3. Stratum granulosum
  Keratohyalin and a hard protein envelope
  form; lamellar bodies release lipids; cells die.

2. Stratum spinosum
  Keratin fibers and lamellar bodies accumulate.

1. Stratum basale
  Cells divide by mitosis and some
  of the newly formed cells become
  the cells of the more superficial strata.

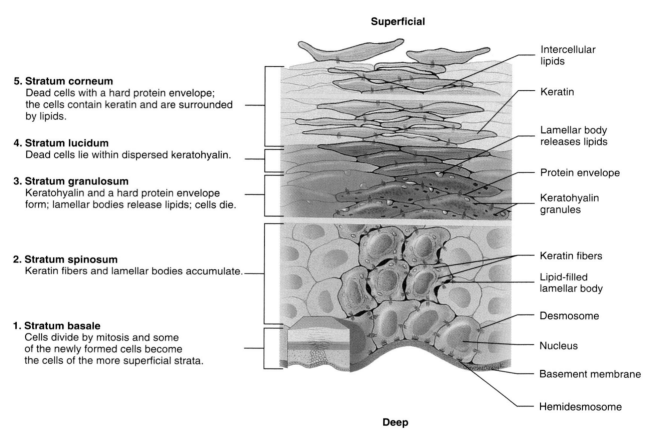

Intercellular
lipids

Keratin

Lamellar body
releases lipids

Protein envelope

Keratohyalin
granules

Keratin fibers

Lipid-filled
lamellar body

Desmosome

Nucleus

Basement membrane

Hemidesmosome

Deep

Epidermal Layers and Keratinization
Figure 5.4

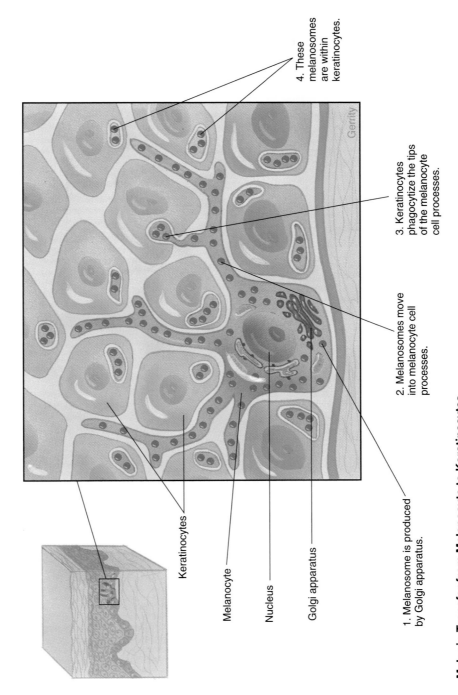

Melanin Transfer from Melanocyte to Keratinocytes
Figure 5.5

4. These melanosomes are within keratinocytes.

3. Keratinocytes phagocytize the tips of the melanocyte cell processes.

2. Melanosomes move into melanocyte cell processes.

Keratinocytes

Melanocyte

Nucleus

Golgi apparatus

1. Melanosome is produced by Golgi apparatus.

Gerrity

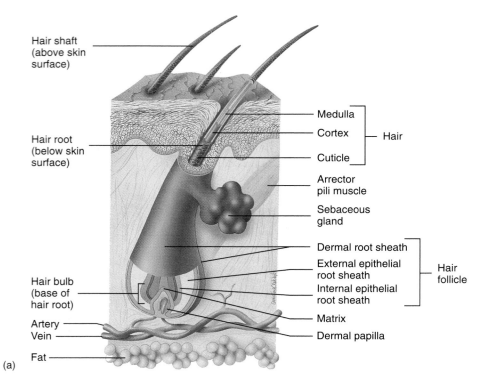

Hair shaft (above skin surface)

Hair root (below skin surface)

Medulla
Cortex ⎤ Hair
Cuticle

Arrector pili muscle

Sebaceous gland

Dermal root sheath
External epithelial root sheath ⎤ Hair follicle
Internal epithelial root sheath

Hair bulb (base of hair root)

Matrix

Dermal papilla

Artery
Vein

Fat

(a)

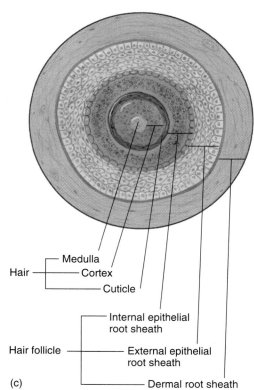

Hair ⎡ Medulla
     ⎢ Cortex
     ⎣ Cuticle

Internal epithelial root sheath

Hair follicle ⎡ External epithelial root sheath
              ⎣ Dermal root sheath

(c)

Hair Follicle
Figure 5.6a, c

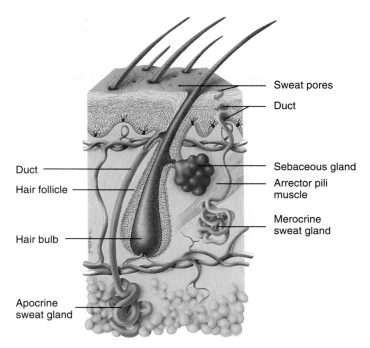

Sweat pores

Duct

Duct

Hair follicle

Hair bulb

Apocrine
sweat gland

Sebaceous gland

Arrector pili
muscle

Merocrine
sweat gland

Glands of the Skin
Figure 5.7

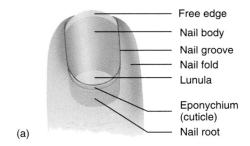

Free edge

Nail body

Nail groove

Nail fold

Lunula

Eponychium
(cuticle)

Nail root

(a)

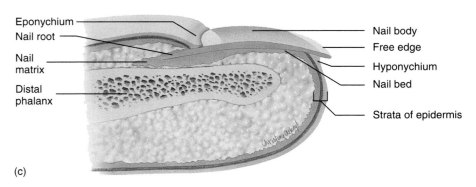

Eponychium

Nail root

Nail
matrix

Distal
phalanx

Nail body

Free edge

Hyponychium

Nail bed

Strata of epidermis

(c)

Nail
Figure 5.8a, c

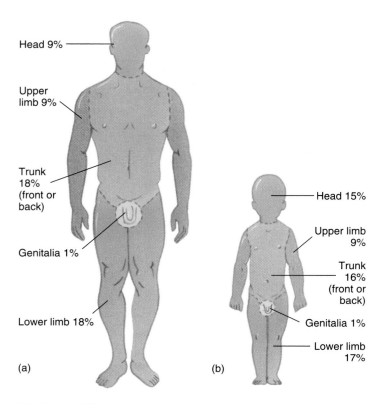

The Rule of Nines
Figure 5B

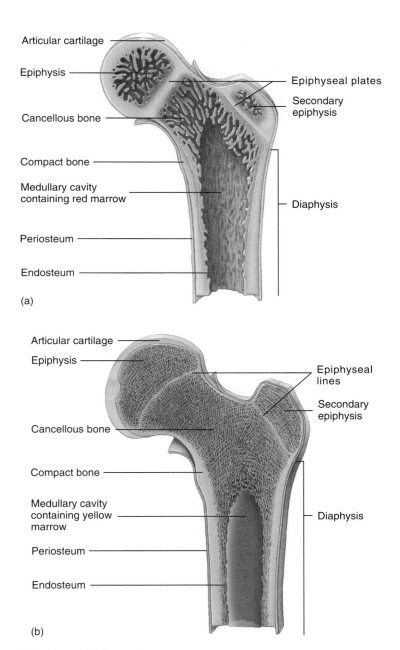

Articular cartilage

Epiphysis

Cancellous bone

Compact bone

Medullary cavity
containing red marrow

Periosteum

Endosteum

(a)

Epiphyseal plates

Secondary
epiphysis

Diaphysis

Articular cartilage

Epiphysis

Cancellous bone

Compact bone

Medullary cavity
containing yellow
marrow

Periosteum

Endosteum

(b)

Epiphyseal
lines

Secondary
epiphysis

Diaphysis

Structure of a Long Bone
Figure 6.3a, b

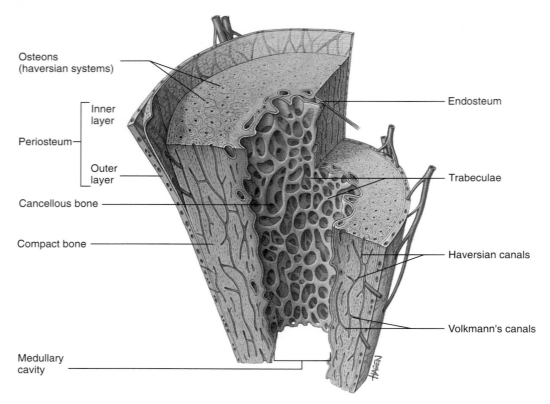

Osteons
(haversian systems)

Periosteum
— Inner layer
— Outer layer

Cancellous bone

Compact bone

Medullary cavity

Endosteum

Trabeculae

Haversian canals

Volkmann's canals

Structure of a Long Bone
Figure 6.3c

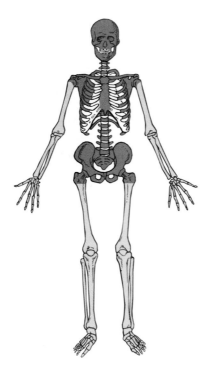

Bone Marrow
Figure 6.4

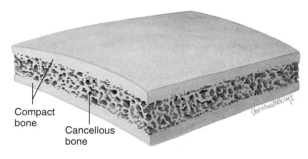

Structure of a Flat Bone
Figure 6.5

Compact bone

Cancellous bone

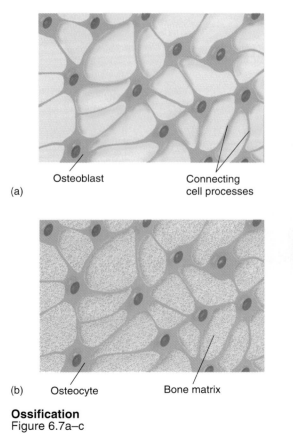

(a)

Osteoblast

Connecting cell processes

(b)

Osteocyte

Bone matrix

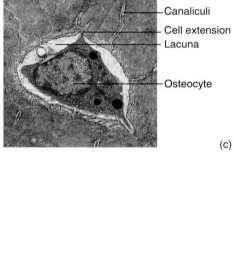

(c)

Canaliculi

Cell extension

Lacuna

Osteocyte

Ossification
Figure 6.7a–c

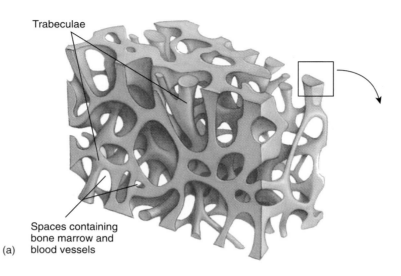

Trabeculae

Spaces containing
bone marrow and
blood vessels

(a)

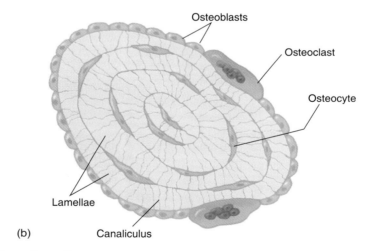

Osteoblasts

Osteoclast

Osteocyte

Lamellae

(b)

Canaliculus

Cancellous Bone
Figure 6.8

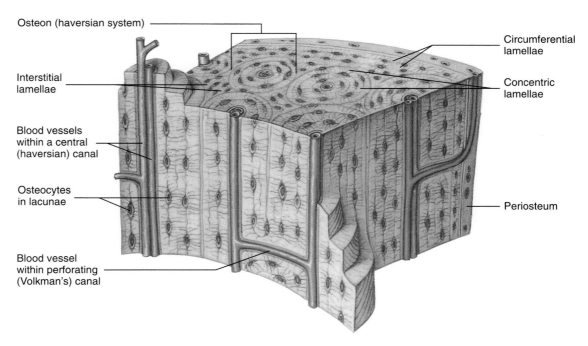

Osteon (haversian system)

Interstitial lamellae

Blood vessels within a central (haversian) canal

Osteocytes in lacunae

Blood vessel within perforating (Volkman's) canal

Circumferential lamellae

Concentric lamellae

Periosteum

Compact Bone
Figure 6.10a

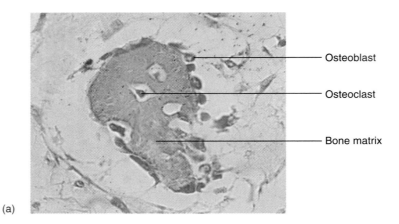

(a)

Osteoblast

Osteoclast

Bone matrix

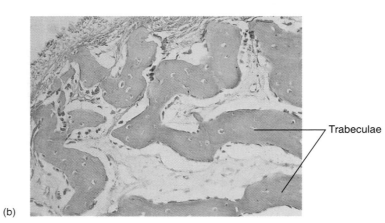

(b)

Trabeculae

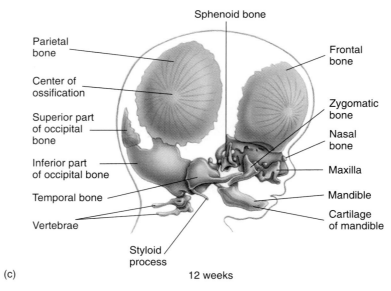

Sphenoid bone

Parietal bone

Frontal bone

Center of ossification

Zygomatic bone

Superior part of occipital bone

Nasal bone

Inferior part of occipital bone

Maxilla

Temporal bone

Mandible

Vertebrae

Cartilage of mandible

Styloid process

(c)

12 weeks

Intramembranous Ossification
Figure 6.11

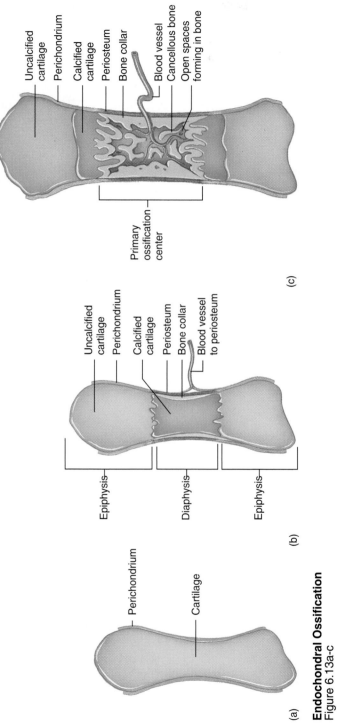

Endochondral Ossification
Figure 6.13a-c

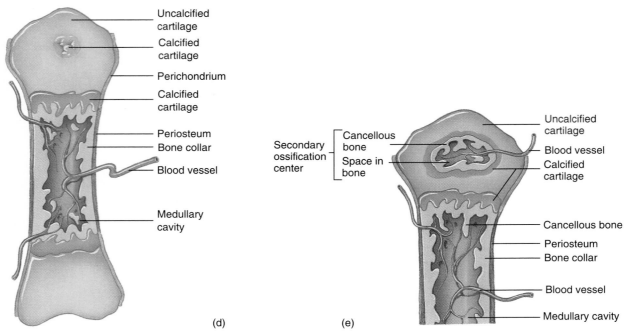

(d)

(e)

Endochondral Ossification
Figure 6.13d, e

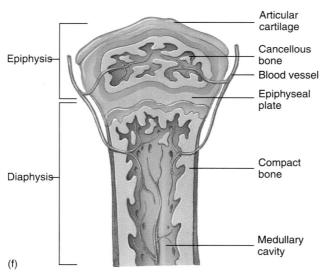

(f)

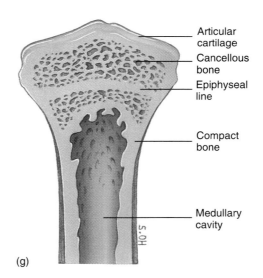

(g)

Endochondral Ossification
Figure 6.13f, g

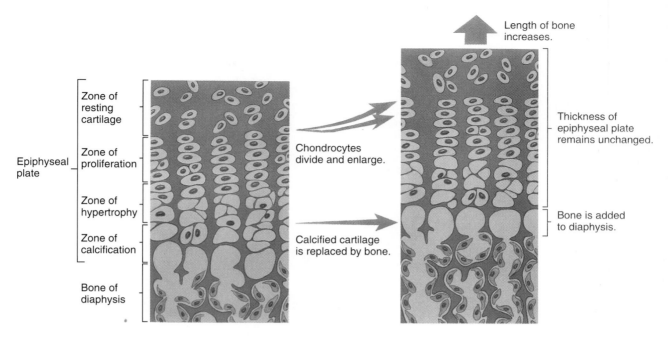

Zone of resting cartilage

Zone of proliferation

Epiphyseal plate

Zone of hypertrophy

Zone of calcification

Bone of diaphysis

Chondrocytes divide and enlarge.

Calcified cartilage is replaced by bone.

Length of bone increases.

Thickness of epiphyseal plate remains unchanged.

Bone is added to diaphysis.

Bone Growth in Length at the Epiphyseal Plate
Figure 6.15

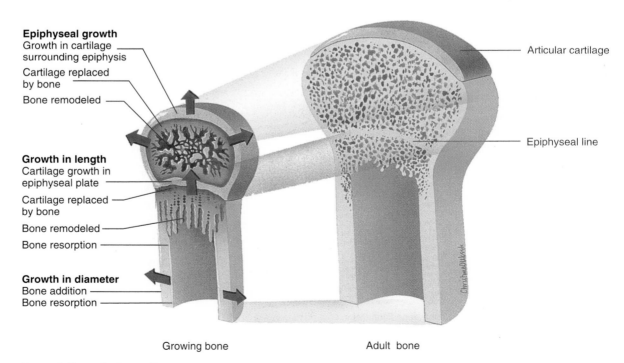

Epiphyseal growth
Growth in cartilage surrounding epiphysis

Cartilage replaced by bone

Bone remodeled

Growth in length
Cartilage growth in epiphyseal plate

Cartilage replaced by bone

Bone remodeled

Bone resorption

Growth in diameter
Bone addition

Bone resorption

Articular cartilage

Epiphyseal line

Growing bone

Adult bone

Remodeling of a Long Bone
Figure 6.17

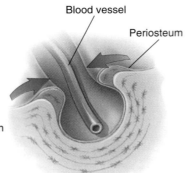

Blood vessel

Periosteum

The surface of a bone consists of grooves and ridges. Blood vessels in the periosteum tend to lie in the grooves. New bone is added to the ridges (arrows), building them up.

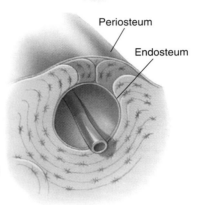

Periosteum

Endosteum

When the bone built on adjacent ridges meets, the groove is transformed into a tunnel. The periosteum of the groove becomes the endosteum of the tunnel.

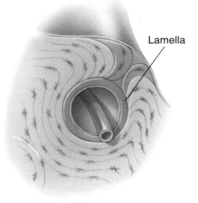

Lamella

Appositional growth by osteoblasts from the endosteum results in the formation of a new lamella.

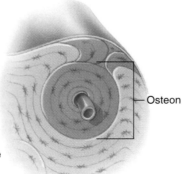

Osteon

Additional bone apposition fills in the tunnel, thus completing the osteon.

Formation of a New Osteon
Figure 6.18

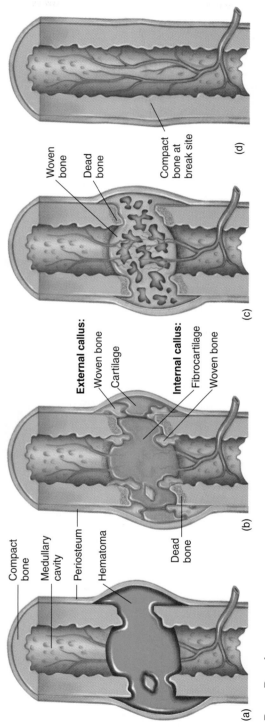

Compact
bone

Medullary
cavity

Periosteum

Hematoma

Dead
bone

External callus:
Woven bone

Cartilage

Internal callus:
Fibrocartilage

Woven bone

Woven
bone

Dead
bone

Compact
bone at
break site

(a)

(b)

(c)

(d)

Bone Repair
Figure 6.19

1. Osteoclasts break down bone and release calcium into the blood, and osteoblasts remove calcium from the blood to make bone. PTH regulates blood calcium levels by indirectly stimulating osteoclast activity, resulting in increased calcium release into the blood. Calcitonin plays a minor role in calcium maintenance by inhibiting osteoclast activity.

2. In the kidneys, PTH also increases calcium reabsorption and promotes the formation of active vitamin D, which increases calcium absorption from the small intestine.

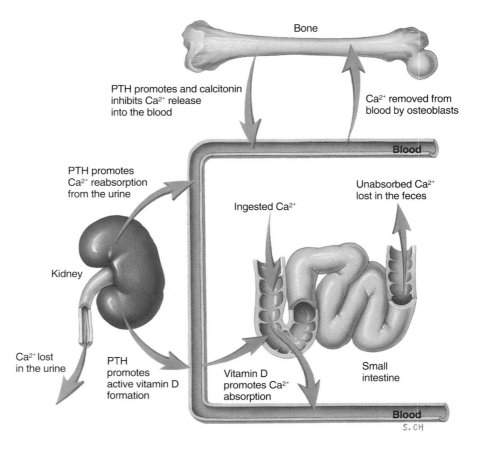

Calcium Homeostasis
Figure 6.20

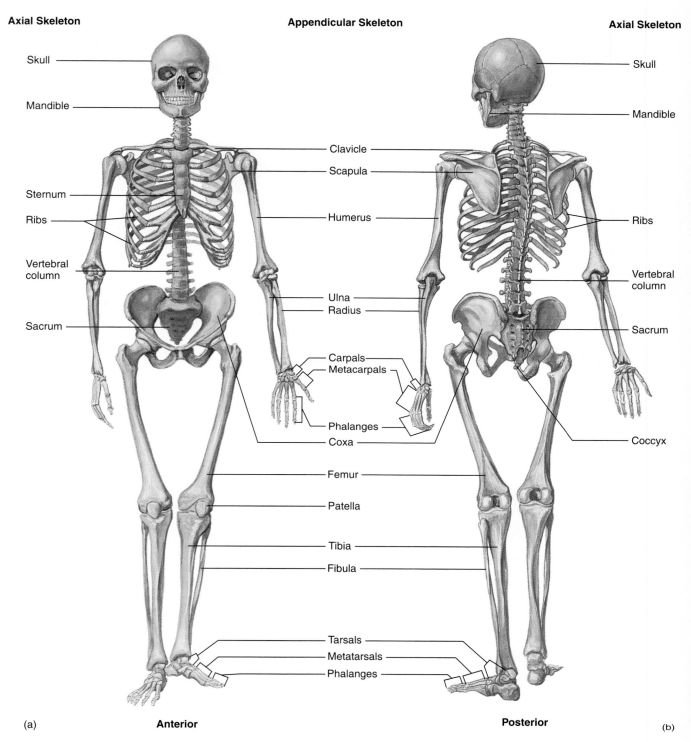

Axial Skeleton

Appendicular Skeleton

Axial Skeleton

Skull

Mandible

Sternum

Ribs

Vertebral
column

Sacrum

Clavicle

Scapula

Humerus

Ulna

Radius

Carpals

Metacarpals

Phalanges

Coxa

Femur

Patella

Tibia

Fibula

Tarsals

Metatarsals

Phalanges

Skull

Mandible

Ribs

Vertebral
column

Sacrum

Coccyx

(a)　　　　　　　　Anterior

Posterior　　　　(b)

The Complete Skeleton (Anterior and Posterior View)
Figure 7.1a, b

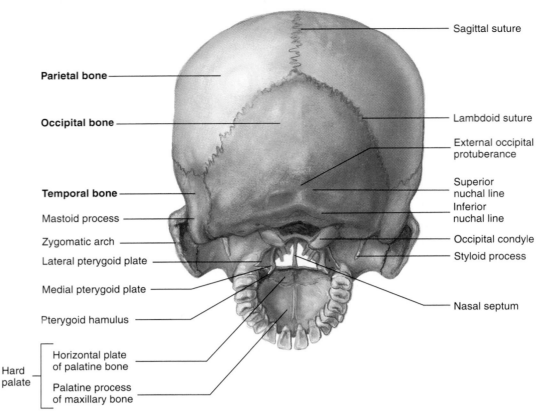

Skull as Seen From Posterior View
Figure 7.4

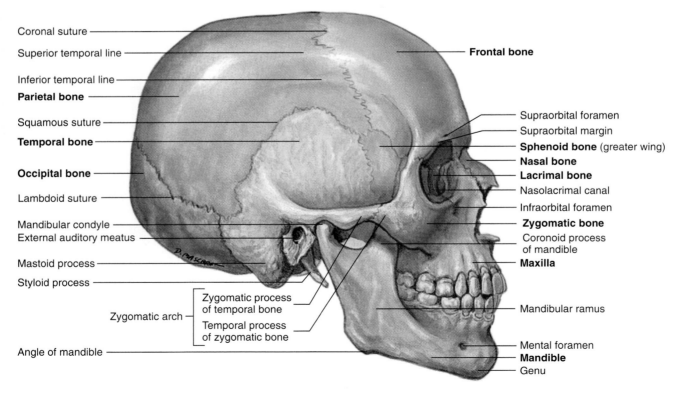

Lateral View of the Skull as Seen from the Right Side
Figure 7.5

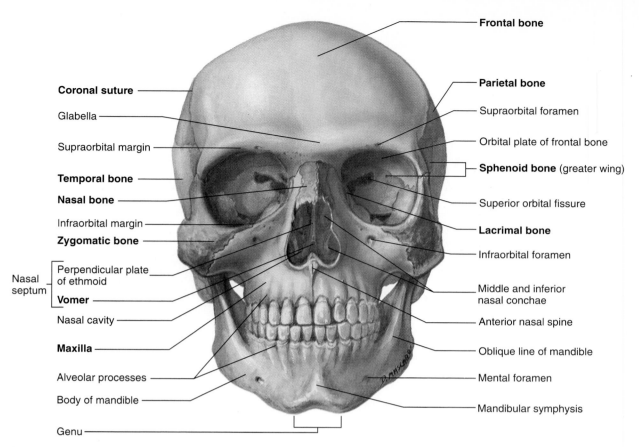

Skull as Seen from the Frontal View
Figure 7.7

Frontal bone

Coronal suture

Glabella

Supraorbital margin

Temporal bone

Nasal bone

Infraorbital margin

Zygomatic bone

Nasal septum {
  Perpendicular plate of ethmoid
  Vomer
}

Nasal cavity

Maxilla

Alveolar processes

Body of mandible

Genu

Parietal bone

Supraorbital foramen

Orbital plate of frontal bone

Sphenoid bone (greater wing)

Superior orbital fissure

Lacrimal bone

Infraorbital foramen

Middle and inferior nasal conchae

Anterior nasal spine

Oblique line of mandible

Mental foramen

Mandibular symphysis

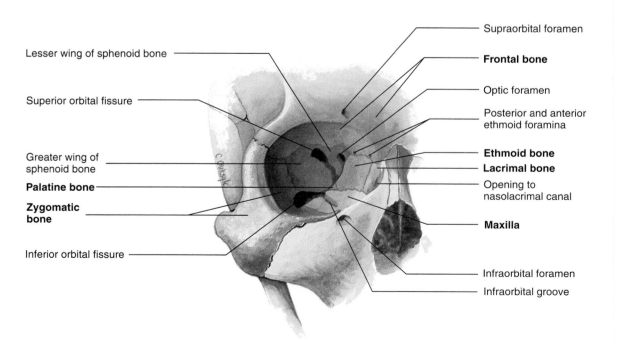

Bones of the Right Orbit
Figure 7.9

Lesser wing of sphenoid bone

Superior orbital fissure

Greater wing of sphenoid bone

Palatine bone

Zygomatic bone

Inferior orbital fissure

Supraorbital foramen

Frontal bone

Optic foramen

Posterior and anterior ethmoid foramina

Ethmoid bone

Lacrimal bone

Opening to nasolacrimal canal

Maxilla

Infraorbital foramen

Infraorbital groove

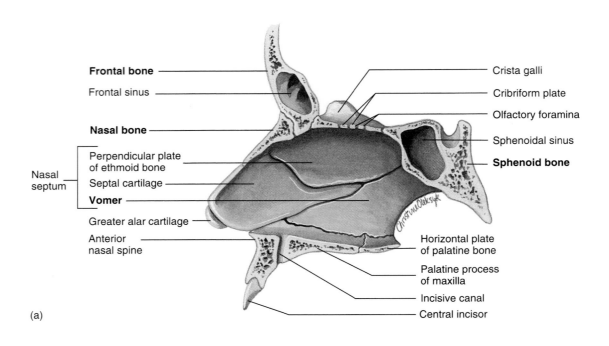

Frontal bone

Frontal sinus

Nasal bone

Nasal septum

Perpendicular plate of ethmoid bone

Septal cartilage

Vomer

Greater alar cartilage

Anterior nasal spine

Crista galli

Cribriform plate

Olfactory foramina

Sphenoidal sinus

Sphenoid bone

Horizontal plate of palatine bone

Palatine process of maxilla

Incisive canal

Central incisor

(a)

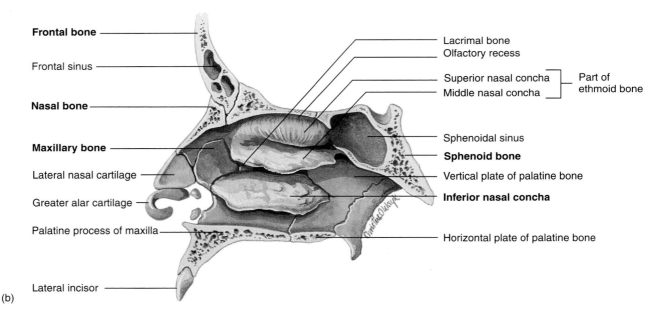

Frontal bone

Frontal sinus

Nasal bone

Maxillary bone

Lateral nasal cartilage

Greater alar cartilage

Palatine process of maxilla

Lateral incisor

Lacrimal bone

Olfactory recess

Superior nasal concha

Middle nasal concha

Part of ethmoid bone

Sphenoidal sinus

Sphenoid bone

Vertical plate of palatine bone

Inferior nasal concha

Horizontal plate of palatine bone

(b)

Bones of the Nasal Cavity
Figure 7.10

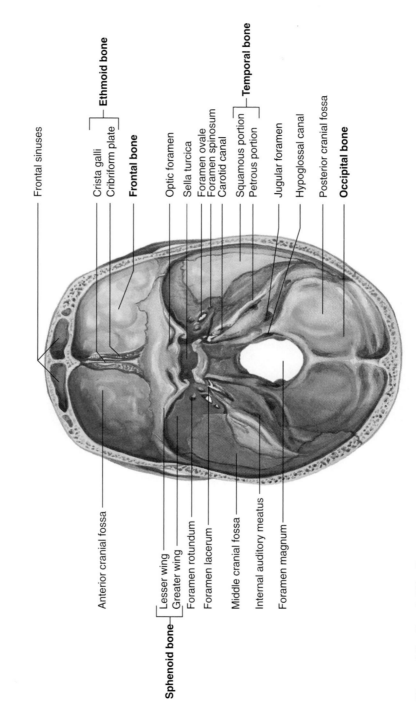

Frontal sinuses

Crista galli
Cribriform plate
Ethmoid bone

Frontal bone

Optic foramen
Sella turcica
Foramen ovale
Foramen spinosum
Carotid canal

Squamous portion
Petrous portion
Temporal bone

Jugular foramen
Hypoglossal canal
Posterior cranial fossa
Occipital bone

Anterior cranial fossa

Lesser wing
Greater wing
Sphenoid bone
Foramen rotundum
Foramen lacerum

Middle cranial fossa
Internal auditory meatus
Foramen magnum

Floor of the Cranial Vault
Figure 7.12

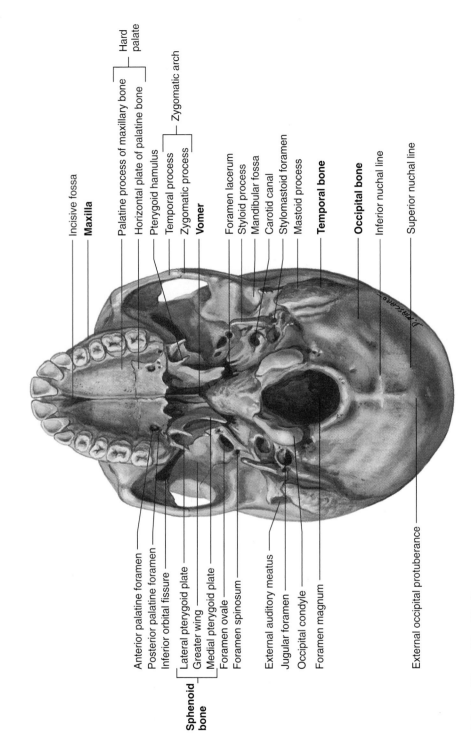

Incisive fossa
Maxilla
Palatine process of maxillary bone — Hard palate
Horizontal plate of palatine bone
Pterygoid hamulus
Temporal process — Zygomatic arch
Zygomatic process
Vomer
Foramen lacerum
Styloid process
Mandibular fossa
Carotid canal
Stylomastoid foramen
Mastoid process
Temporal bone
Occipital bone
Inferior nuchal line
Superior nuchal line

Sphenoid bone
Anterior palatine foramen
Posterior palatine foramen
Inferior orbital fissure
Lateral pterygoid plate
Greater wing
Medial pterygoid plate
Foramen ovale
Foramen spinosum
External auditory meatus
Jugular foramen
Occipital condyle
Foramen magnum
External occipital protuberance

Inferior View of the Skull
Figure 7.13

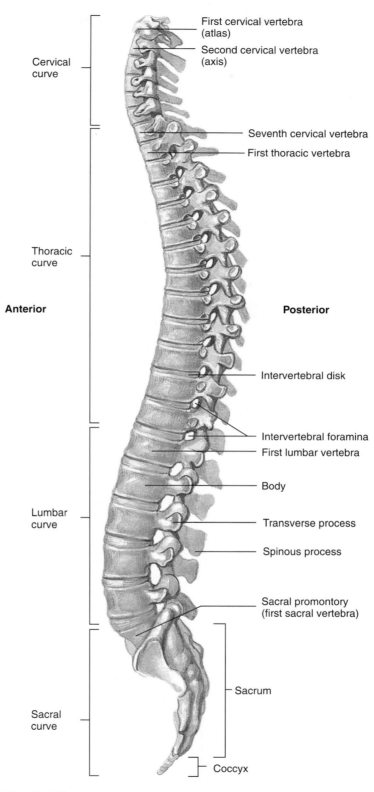

First cervical vertebra (atlas)

Second cervical vertebra (axis)

Cervical curve

Seventh cervical vertebra

First thoracic vertebra

Thoracic curve

Anterior

Posterior

Intervertebral disk

Intervertebral foramina

First lumbar vertebra

Body

Lumbar curve

Transverse process

Spinous process

Sacral promontory (first sacral vertebra)

Sacrum

Sacral curve

Coccyx

Vertebral Column
Figure 7.14

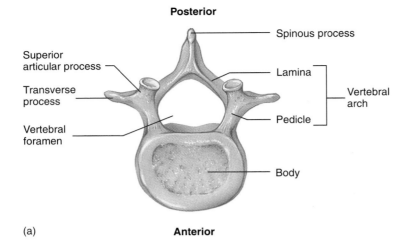

Posterior

Spinous process

Superior
articular process

Lamina

Vertebral
arch

Transverse
process

Pedicle

Vertebral
foramen

Body

(a)

Anterior

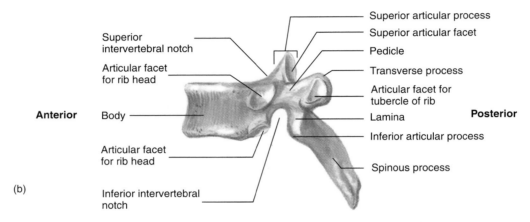

Superior articular process

Superior articular facet

Superior
intervertebral notch

Pedicle

Articular facet
for rib head

Transverse process

Articular facet for
tubercle of rib

Anterior

Body

Lamina

Posterior

Inferior articular process

Articular facet
for rib head

Spinous process

(b)

Inferior intervertebral
notch

Vertebra
Figure 7.16a, b

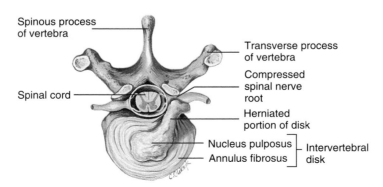

Spinous process
of vertebra

Transverse process
of vertebra

Compressed
spinal nerve
root

Spinal cord

Herniated
portion of disk

Nucleus pulposus

Annulus fibrosus

Intervertebral
disk

Herniated Disk
Figure 7.17a

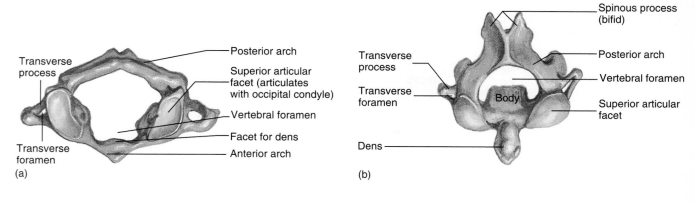

Transverse process

Posterior arch

Superior articular facet (articulates with occipital condyle)

Vertebral foramen

Facet for dens

Transverse foramen

Anterior arch

(a)

Spinous process (bifid)

Transverse process

Transverse foramen

Posterior arch

Vertebral foramen

Body

Superior articular facet

Dens

(b)

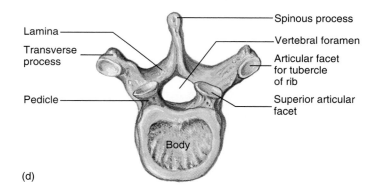

Lamina

Pedicle

Transverse foramen

Transverse process

Spinous process (bifid)

Vertebral foramen

Superior articular facet

Body

(c)

Vertebrae
Figure 7.20a-c

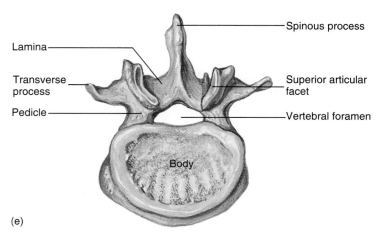

Lamina

Transverse process

Pedicle

Spinous process

Vertebral foramen

Articular facet for tubercle of rib

Superior articular facet

Body

(d)

Lamina

Transverse process

Pedicle

Spinous process

Superior articular facet

Vertebral foramen

Body

(e)

Vertebrae
Figure 7.20d, e

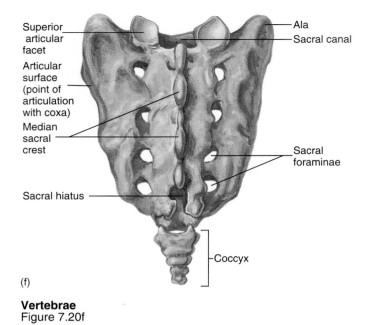

Superior articular facet

Articular surface (point of articulation with coxa)

Median sacral crest

Sacral hiatus

Ala

Sacral canal

Sacral foraminae

Coccyx

(f)

Vertebrae
Figure 7.20f

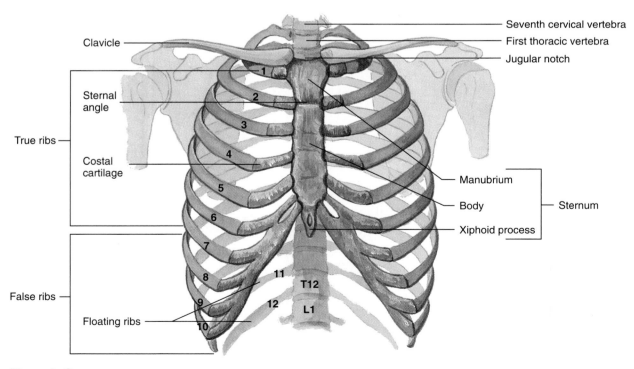

Clavicle

Sternal angle

Costal cartilage

True ribs

False ribs

Floating ribs

Seventh cervical vertebra

First thoracic vertebra

Jugular notch

Manubrium

Body

Xiphoid process

Sternum

1
2
3
4
5
6
7
8
9
10
11
12

T12

L1

Thoracic Cage
Figure 7.21a

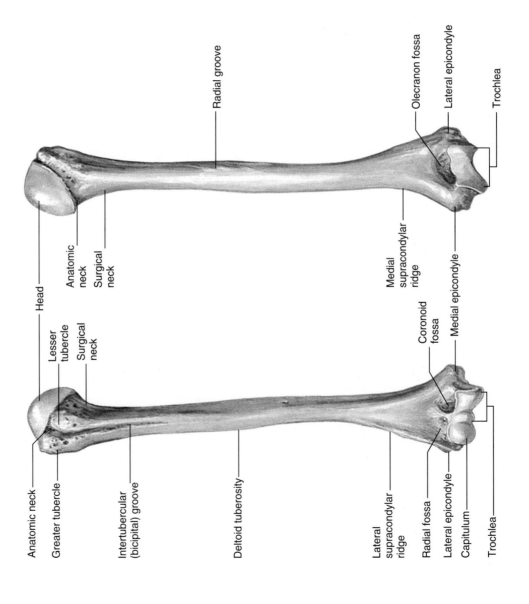

Radial groove

Olecranon fossa

Lateral epicondyle

Trochlea

Anatomic neck

Surgical neck

Head

Medial supracondylar ridge

Medial epicondyle

Lesser tubercle

Surgical neck

Coronoid fossa

Anatomic neck

Greater tubercle

Intertubercular (bicipital) groove

Deltoid tuberosity

Lateral supracondylar ridge

Radial fossa

Lateral epicondyle

Capitulum

Trochlea

Right Humerus
Figure 7.24

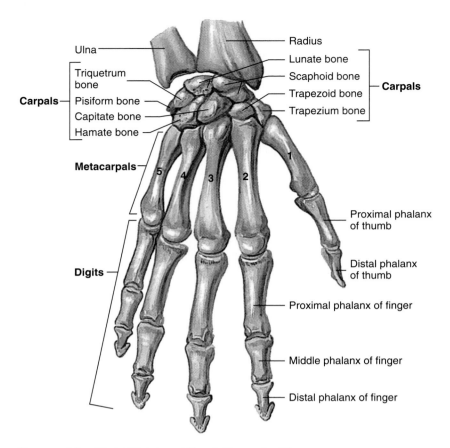

Ulna

Radius

Triquetrum
bone

Lunate bone

Scaphoid bone

Carpals

Pisiform bone

Trapezoid bone

Carpals

Capitate bone

Trapezium bone

Hamate bone

Metacarpals

5   4   3   2   1

Proximal phalanx
of thumb

Digits

Distal phalanx
of thumb

Proximal phalanx of finger

Middle phalanx of finger

Distal phalanx of finger

Bones of the Right Wrist and Hand, Posterior View
Figure 7.27

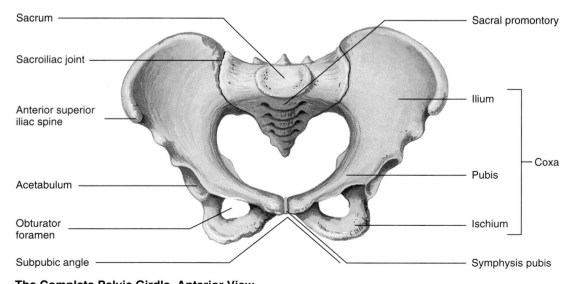

Sacrum

Sacral promontory

Sacroiliac joint

Anterior superior
iliac spine

Ilium

Coxa

Acetabulum

Pubis

Obturator
foramen

Ischium

Subpubic angle

Symphysis pubis

The Complete Pelvic Girdle, Anterior View
Figure 7.28

92

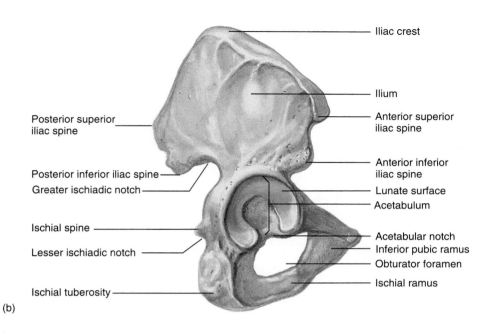

Iliac crest

Ilium

Posterior superior
iliac spine

Anterior superior
iliac spine

Posterior inferior iliac spine

Anterior inferior
iliac spine

Greater ischiadic notch

Lunate surface

Acetabulum

Ischial spine

Acetabular notch

Inferior pubic ramus

Lesser ischiadic notch

Obturator foramen

Ischial ramus

Ischial tuberosity

(b)

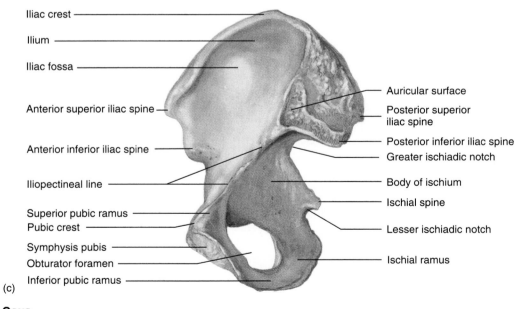

Iliac crest

Ilium

Iliac fossa

Anterior superior iliac spine

Auricular surface

Posterior superior
iliac spine

Anterior inferior iliac spine

Posterior inferior iliac spine

Greater ischiadic notch

Iliopectineal line

Body of ischium

Superior pubic ramus

Ischial spine

Pubic crest

Lesser ischiadic notch

Symphysis pubis

Obturator foramen

Ischial ramus

Inferior pubic ramus

(c)

Coxa
Figure 7.29b, c

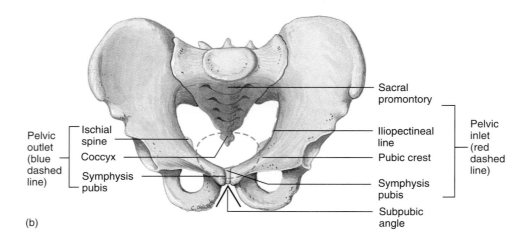

(b)

Pelvic outlet (blue dashed line) — Ischial spine — Coccyx — Symphysis pubis

Sacral promontory

Iliopectineal line

Pubic crest

Symphysis pubis

Subpubic angle

Pelvic inlet (red dashed line)

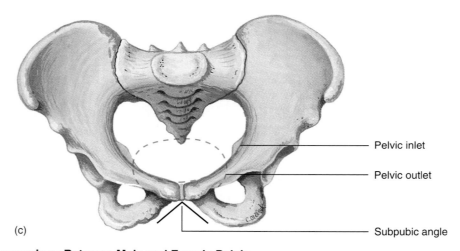

(c)

Pelvic inlet

Pelvic outlet

Subpubic angle

Comparison Between Male and Female Pelvis
Figure 7.31b, c

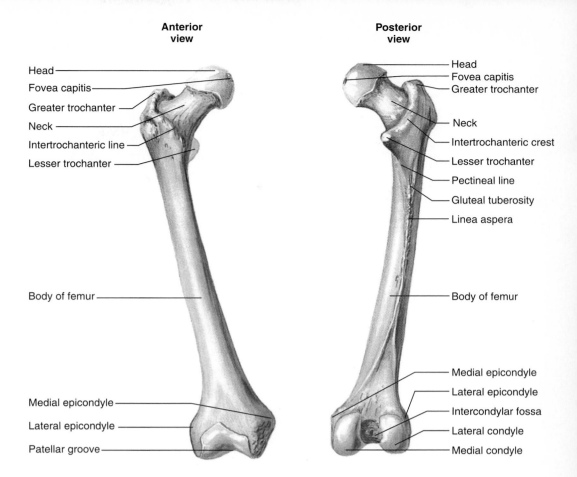

Anterior view

Head

Fovea capitis

Greater trochanter

Neck

Intertrochanteric line

Lesser trochanter

Body of femur

Medial epicondyle

Lateral epicondyle

Patellar groove

Posterior view

Head

Fovea capitis

Greater trochanter

Neck

Intertrochanteric crest

Lesser trochanter

Pectineal line

Gluteal tuberosity

Linea aspera

Body of femur

Medial epicondyle

Lateral epicondyle

Intercondylar fossa

Lateral condyle

Medial condyle

Right Femur
Figure 7.32

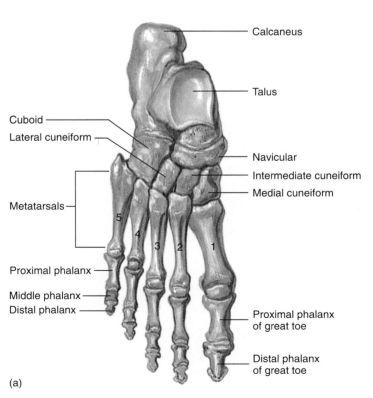

Calcaneus

Talus

Cuboid

Lateral cuneiform

Navicular

Intermediate cuneiform

Medial cuneiform

Metatarsals

5

4

3

2

1

Proximal phalanx

Middle phalanx

Distal phalanx

Proximal phalanx
of great toe

Distal phalanx
of great toe

(a)

Bones of the Right Ankle and Foot
Figure 7.36, *(Continued)*

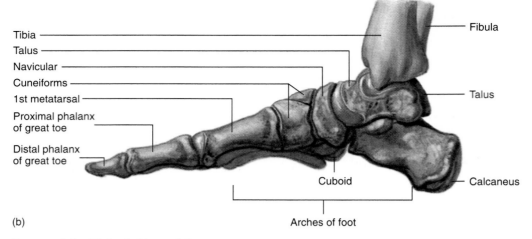

Tibia

Talus

Navicular

Cuneiforms

1st metatarsal

Proximal phalanx
of great toe

Distal phalanx
of great toe

Fibula

Talus

Calcaneus

Cuboid

Arches of foot

(b)

Bones of the Right Ankle and Foot
Figure 7.36

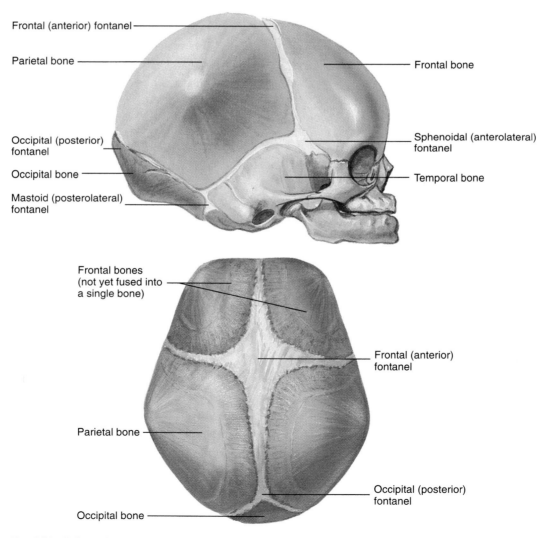

Frontal (anterior) fontanel

Parietal bone

Occipital (posterior)
fontanel

Occipital bone

Mastoid (posterolateral)
fontanel

Frontal bone

Sphenoidal (anterolateral)
fontanel

Temporal bone

Frontal bones
(not yet fused into
a single bone)

Frontal (anterior)
fontanel

Parietal bone

Occipital (posterior)
fontanel

Occipital bone

Fetal Skull Showing Fontanels
Figure 8.1

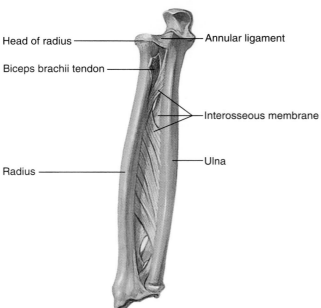

Head of radius

Biceps brachii tendon

Radius

Annular ligament

Interosseous membrane

Ulna

Right Radioulnar Syndesmosis (Interosseous Membrane)
Figure 8.2

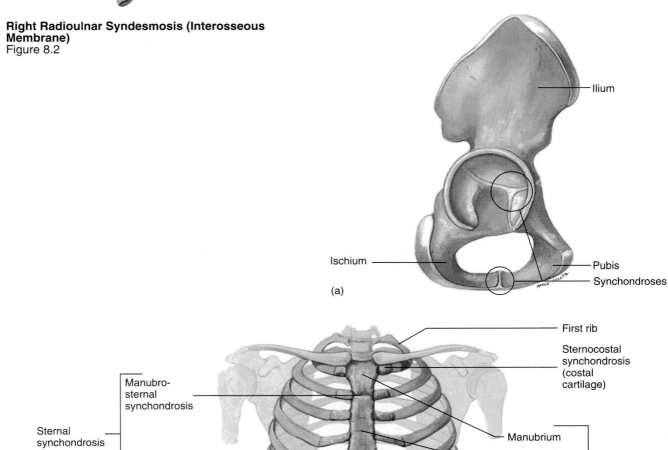

Ilium

Ischium

Pubis

Synchondroses

(a)

First rib

Sternocostal synchondrosis (costal cartilage)

Manubro-sternal synchondrosis

Sternal synchondrosis

Xiphosternal synchondrosis

Manubrium

Body

Xiphoid process

Sternum

(b)

Synchondroses
Figure 8.3a, b

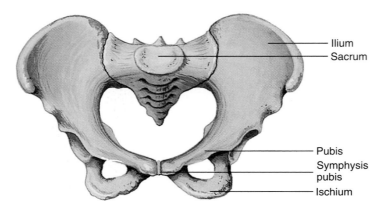

Symphysis Pubis
Figure 8.4

Ilium
Sacrum

Pubis
Symphysis pubis
Ischium

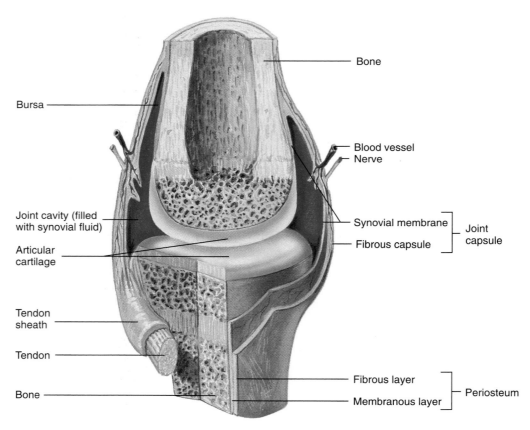

Bone

Bursa

Blood vessel
Nerve

Joint cavity (filled with synovial fluid)

Synovial membrane
Fibrous capsule

Joint capsule

Articular cartilage

Tendon sheath

Tendon

Bone

Fibrous layer

Membranous layer

Periosteum

Structure of a Synovial Joint
Figure 8.5

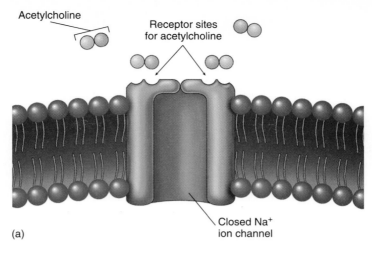

Acetylcholine

Receptor sites
for acetylcholine

Closed Na⁺
ion channel

(a)

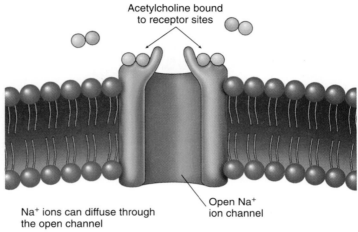

Acetylcholine bound
to receptor sites

Na⁺ ions can diffuse through
the open channel

Open Na⁺
ion channel

(b)

Membrane-Bound Receptors Directly Affecting Membrane Permeability
Figure 9.3

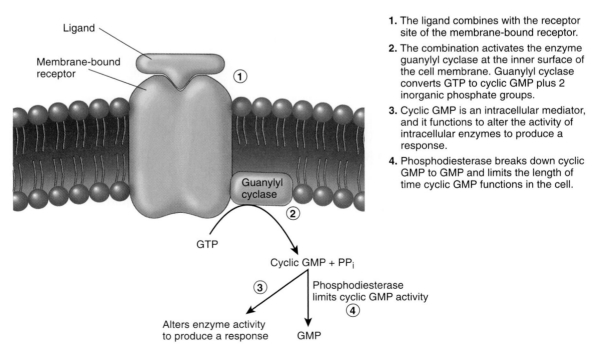

Ligand

Membrane-bound
receptor

①

Guanylyl
cyclase

②

GTP

Cyclic GMP + PP$_i$

③

Phosphodiesterase
limits cyclic GMP activity

④

Alters enzyme activity
to produce a response

GMP

1. The ligand combines with the receptor site of the membrane-bound receptor.

2. The combination activates the enzyme guanylyl cyclase at the inner surface of the cell membrane. Guanylyl cyclase converts GTP to cyclic GMP plus 2 inorganic phosphate groups.

3. Cyclic GMP is an intracellular mediator, and it functions to alter the activity of intracellular enzymes to produce a response.

4. Phosphodiesterase breaks down cyclic GMP to GMP and limits the length of time cyclic GMP functions in the cell.

Membrane Receptors that Increase Cyclic GMP Synthesis
Figure 9.4

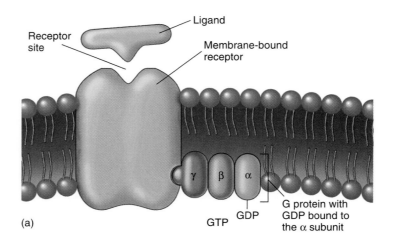

(a)

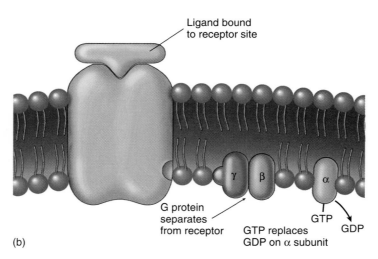

(b)

Membrane-Bound Receptors and G Proteins
Figure 9.5

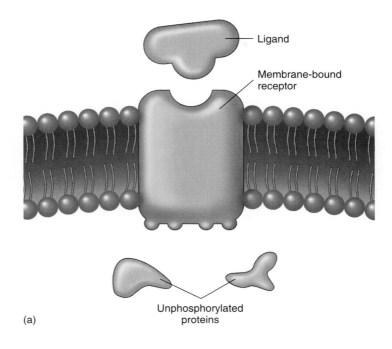

(a)

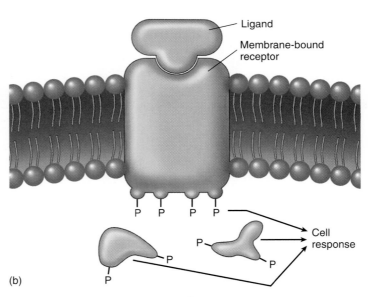

(b)

Receptors That Phosphorylate Proteins
Figure 9.9

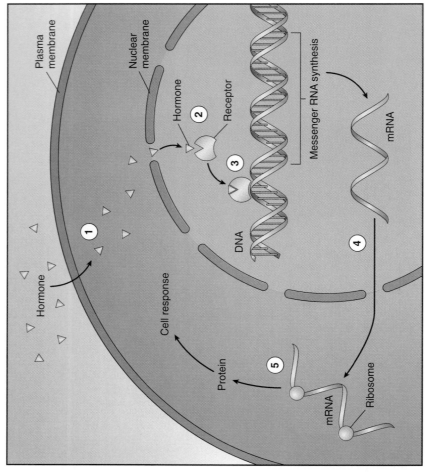

1. The ligand diffuses through the plasma membrane and enters the cytoplasm of the cell.

2. The ligand combines with the receptor in the nucleus (or in the cytoplasm).

3. The receptor with the ligand bound to it interacts with DNA and increases the synthesis of specific messenger RNA molecules.

4. The messenger RNA passes from the nucleus to the cytoplasm.

5. In the cytoplasm of the cell, messenger RNA combines with ribosomes. New protein molecules, which produce the response of the cell to the ligand, are synthesized.

Actions of Intracellular Receptors
Figure 9.10

102

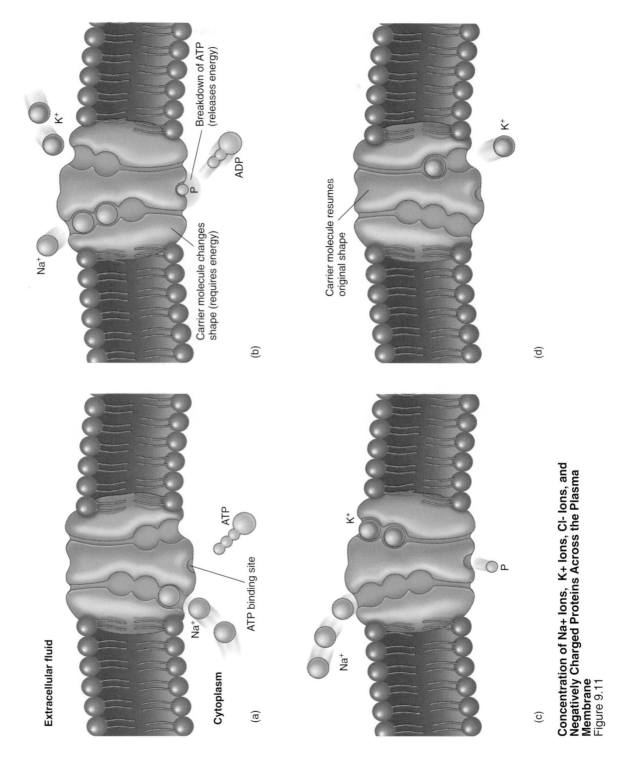

Concentration of Na+ Ions, K+ Ions, Cl- Ions, and Negatively Charged Proteins Across the Plasma Membrane
Figure 9.11

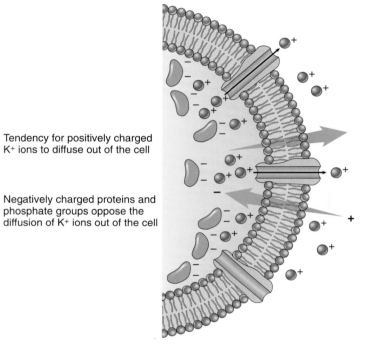

Tendency for positively charged
K+ ions to diffuse out of the cell

Negatively charged proteins and
phosphate groups oppose the
diffusion of K+ ions out of the cell

Potassium Ions and the Resting Membrane Potential
Figure 9.14

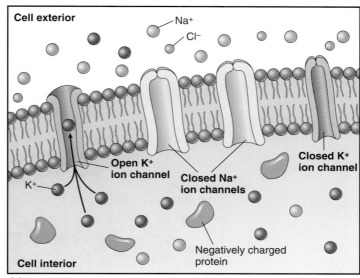

(a)

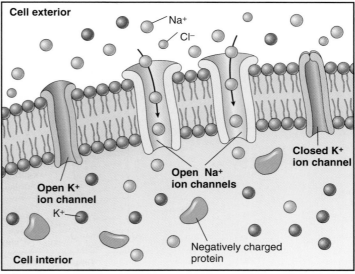

(b)

Stimuli and Membrane Permeability
Figure 9.16

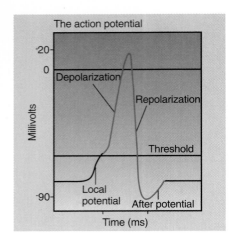

The action potential

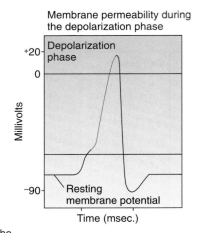

Membrane permeability during the depolarization phase

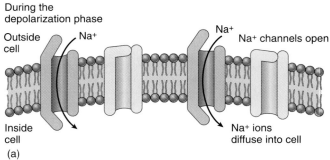

During the depolarization phase

Outside cell

Na⁺   Na⁺   Na⁺ channels open

Inside cell

Na⁺ ions diffuse into cell

(a)

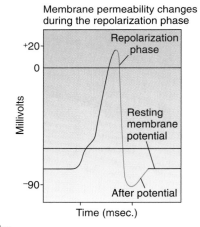

Membrane permeability changes during the repolarization phase

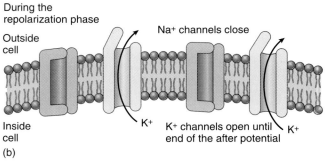

During the repolarization phase

Outside cell

Na⁺ channels close

Inside cell

K⁺   K⁺ channels open until end of the after potential   K⁺

(b)

The Action Potential and Permeability Changes During the Action Potential
Figures 9.18, 9.20

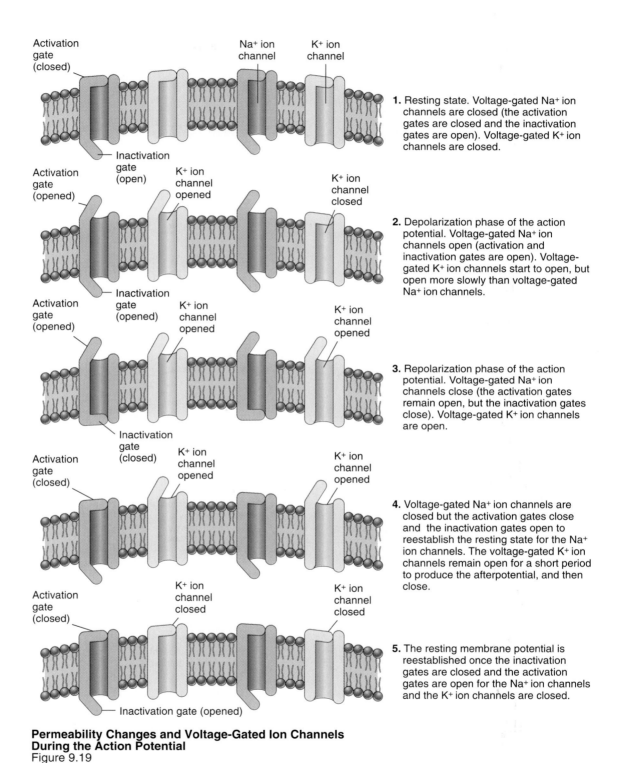

Activation gate (closed)

Na+ ion channel

K+ ion channel

Inactivation gate (open)

1. Resting state. Voltage-gated Na+ ion channels are closed (the activation gates are closed and the inactivation gates are open). Voltage-gated K+ ion channels are closed.

Activation gate (opened)

K+ ion channel opened

K+ ion channel closed

Inactivation gate (opened)

2. Depolarization phase of the action potential. Voltage-gated Na+ ion channels open (activation and inactivation gates are open). Voltage-gated K+ ion channels start to open, but open more slowly than voltage-gated Na+ ion channels.

Activation gate (opened)

K+ ion channel opened

K+ ion channel opened

Inactivation gate (closed)

3. Repolarization phase of the action potential. Voltage-gated Na+ ion channels close (the activation gates remain open, but the inactivation gates close). Voltage-gated K+ ion channels are open.

Activation gate (closed)

K+ ion channel opened

K+ ion channel opened

4. Voltage-gated Na+ ion channels are closed but the activation gates close and the inactivation gates open to reestablish the resting state for the Na+ ion channels. The voltage-gated K+ ion channels remain open for a short period to produce the afterpotential, and then close.

Activation gate (closed)

K+ ion channel closed

K+ ion channel closed

Inactivation gate (opened)

5. The resting membrane potential is reestablished once the inactivation gates are closed and the activation gates are open for the Na+ ion channels and the K+ ion channels are closed.

Permeability Changes and Voltage-Gated Ion Channels During the Action Potential
Figure 9.19

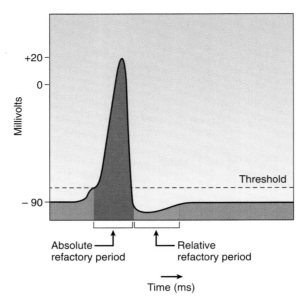

The Refractory Period
Figure 9.21

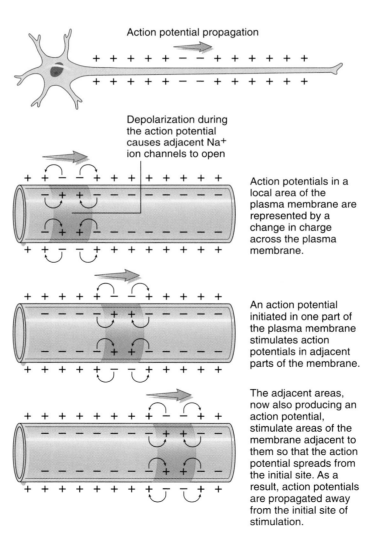

Action potential propagation

Depolarization during the action potential causes adjacent Na⁺ ion channels to open

Action potentials in a local area of the plasma membrane are represented by a change in charge across the plasma membrane.

An action potential initiated in one part of the plasma membrane stimulates action potentials in adjacent parts of the membrane.

The adjacent areas, now also producing an action potential, stimulate areas of the membrane adjacent to them so that the action potential spreads from the initial site. As a result, action potentials are propagated away from the initial site of stimulation.

Action Potential Propagation
Figure 9.22

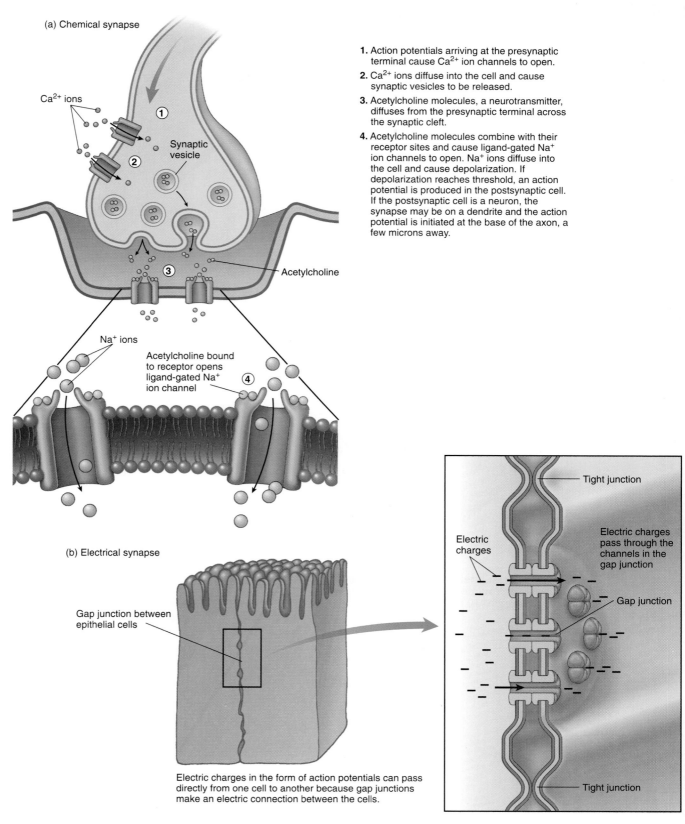

(a) Chemical synapse

Ca²⁺ ions

①

②

Synaptic vesicle

③

Acetylcholine

Na⁺ ions

Acetylcholine bound to receptor opens ligand-gated Na⁺ ion channel

④

1. Action potentials arriving at the presynaptic terminal cause Ca^{2+} ion channels to open.
2. Ca^{2+} ions diffuse into the cell and cause synaptic vesicles to be released.
3. Acetylcholine molecules, a neurotransmitter, diffuses from the presynaptic terminal across the synaptic cleft.
4. Acetylcholine molecules combine with their receptor sites and cause ligand-gated Na^+ ion channels to open. Na^+ ions diffuse into the cell and cause depolarization. If depolarization reaches threshold, an action potential is produced in the postsynaptic cell. If the postsynaptic cell is a neuron, the synapse may be on a dendrite and the action potential is initiated at the base of the axon, a few microns away.

(b) Electrical synapse

Gap junction between epithelial cells

Electric charges in the form of action potentials can pass directly from one cell to another because gap junctions make an electric connection between the cells.

Tight junction

Electric charges pass through the channels in the gap junction

Electric charges

Gap junction

Tight junction

Synapses
Figure 9.23

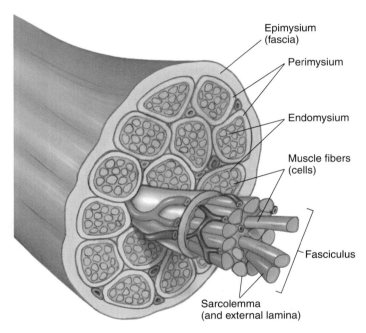

Relationship Between Muscle Fiber, Fasciculi, and Associated Connective Tissues
Figure 10.2

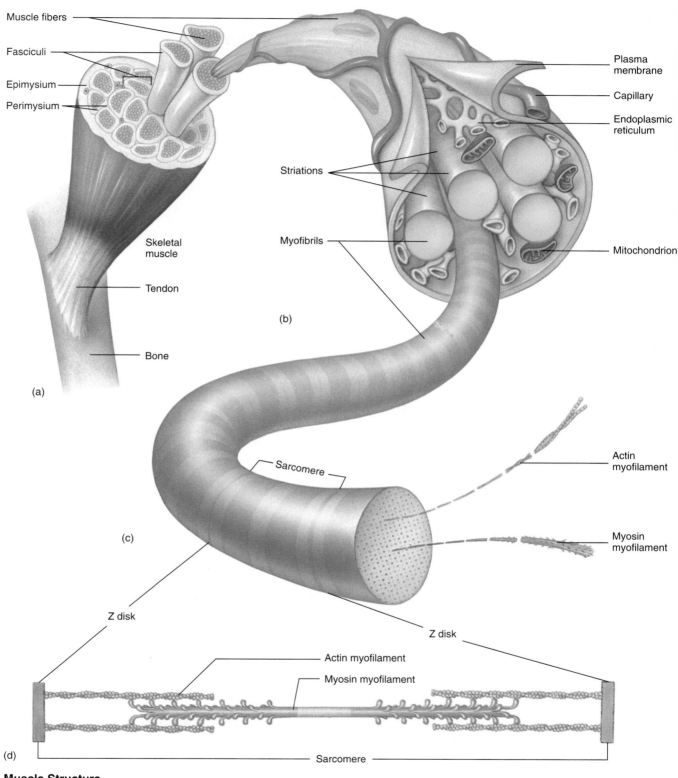

Muscle fibers

Fasciculi

Epimysium

Perimysium

Plasma membrane

Capillary

Endoplasmic reticulum

Striations

Myofibrils

Mitochondrion

Skeletal muscle

Tendon

Bone

(a)

(b)

(c)

Sarcomere

Actin myofilament

Myosin myofilament

Z disk

Z disk

Actin myofilament

Myosin myofilament

(d)

Sarcomere

Muscle Structure
Figure 10.3

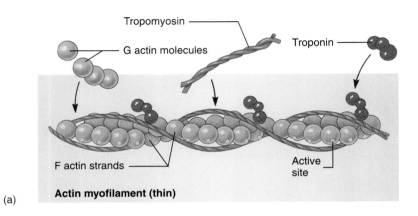

(a) **Actin myofilament (thin)**

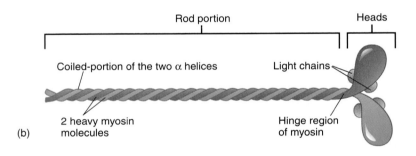

(b)

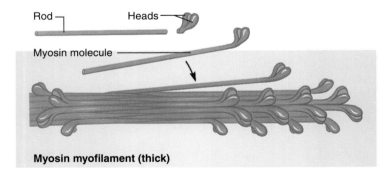

(c) **Myosin myofilament (thick)**

Structure of Actin and Myosin
Figure 10.6

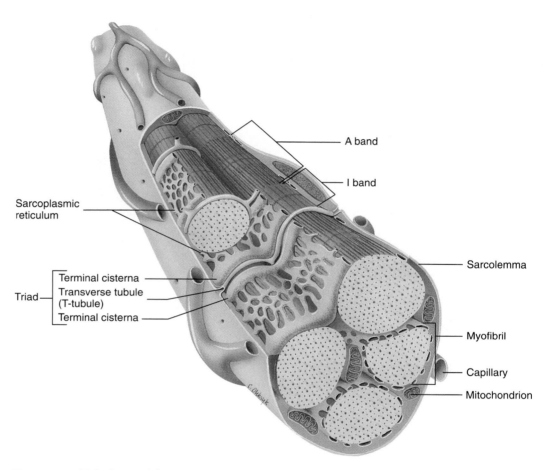

A band

I band

Sarcoplasmic
reticulum

Sarcolemma

Terminal cisterna

Transverse tubule
(T-tubule)

Triad

Terminal cisterna

Myofibril

Capillary

Mitochondrion

Transverse Tubules and Sarcoplasmic Reticulum
Figure 10.7

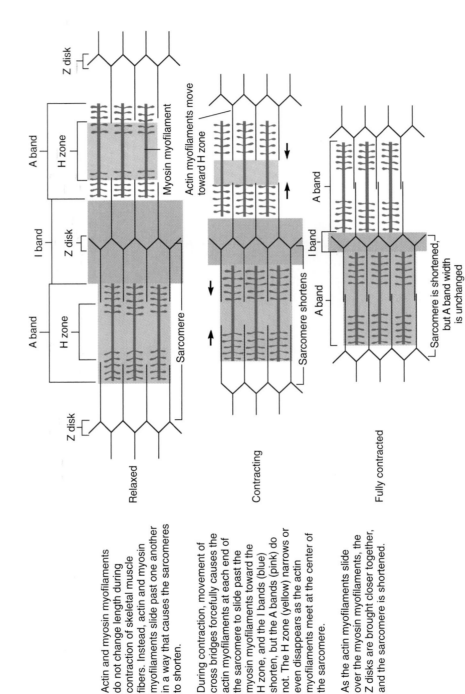

Actin and myosin myofilaments do not change length during contraction of skeletal muscle fibers. Instead, actin and myosin myofilaments slide past one another in a way that causes the sarcomeres to shorten.

During contraction, movement of cross bridges forcefully causes the actin myofilaments at each end of the sarcomere to slide past the myosin myofilaments toward the H zone, and the I bands (blue) shorten, but the A bands (pink) do not. The H zone (yellow) narrows or even disappears as the actin myofilaments meet at the center of the sarcomere.

As the actin myofilaments slide over the myosin myofilaments, the Z disks are brought closer together, and the sarcomere is shortened.

Sarcomere Shortening, Sliding Filament Model
Figure 10.8

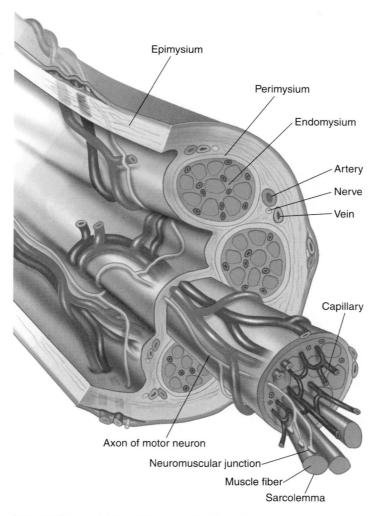

Epimysium

Perimysium

Endomysium

Artery

Nerve

Vein

Capillary

Axon of motor neuron

Neuromuscular junction

Muscle fiber

Sarcolemma

Innervation and Blood Supply of a Muscle
Figure 10.9

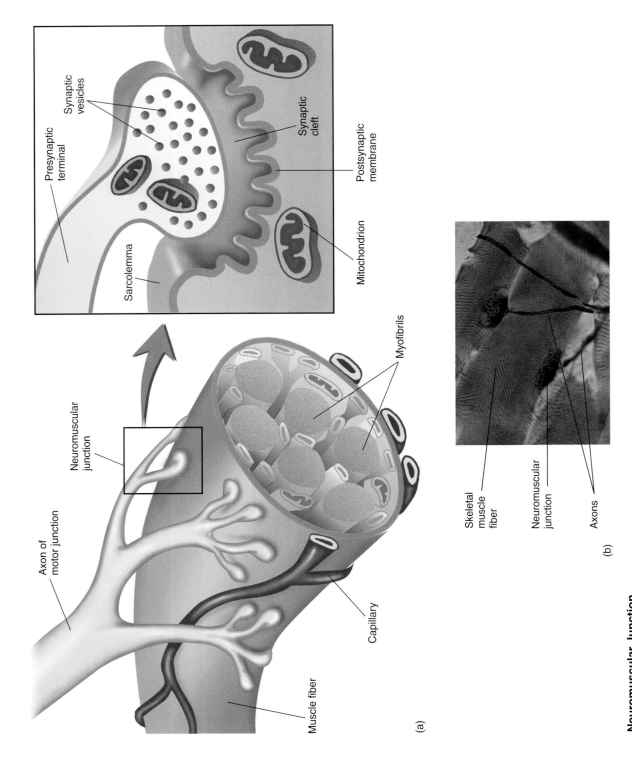

Synaptic vesicles

Presynaptic terminal

Synaptic cleft

Postsynaptic membrane

Mitochondrion

Sarcolemma

Myofibrils

Neuromuscular junction

Axon of motor junction

Capillary

Muscle fiber

(a)

Skeletal muscle fiber

Neuromuscular junction

Axons

(b)

Neuromuscular Junction
Figure 10.10

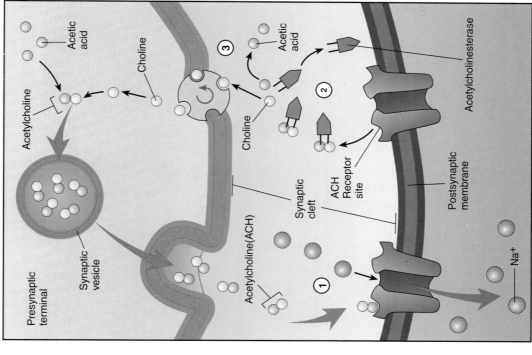

1. Once acetylcholine is released into the synaptic cleft it binds to the receptors for acetylcholine on the postsynaptic membrane and causes Na+ ion channels to open.

2. Acetylcholine is rapidly broken down in the synaptic cleft by acetylcholinesterase to acetic acid and choline.

3. The choline is reabsorbed by the presynaptic terminal and combined with acetic acid to form more acetylcholine, which enters synaptic vesicles. Acetic acid is taken up by many cell types

(b)

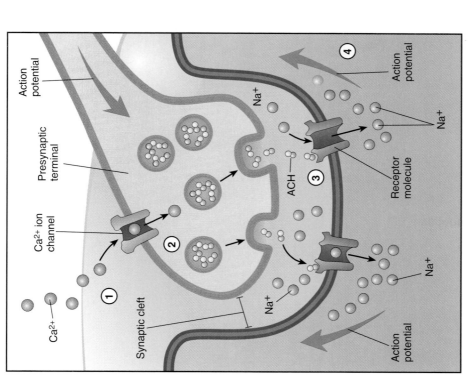

1. An action potential arrives at the presynaptic terminal causing voltage-gated Ca2+ ion channels to open, increasing the Ca2+ ion permeability of the presynaptic terminal.

2. Ca2+ ions enter the presynaptic terminal and initiate the release of a neurotransmitter, acetylcholine (ACH), from synaptic vesicles in the presynaptic terminal.

3. Diffusion of ACH across the synaptic cleft and binding of ACH to ACH receptors on the postsynaptic muscle fiber membrane causes an increase in the permeability of ligand-gated Na+ ion channels.

4. The increase in Na+ ion permeability results in depolarization of the postsynaptic membrane; once threshold has been reached a postsynaptic action potential results.

(a)

Function of the Neuromuscular Junction
Figure 10.11

(a) An action potential is propagated along the sarcolemma of the skeletal muscle, causing a depolarization to spread along the membrane of the T tubules.

(b) The depolarization of the T tubule causes voltage-gated Ca^{2+} ion channels to open, resulting in an increase in the permeability of the sarcoplasmic reticulum to Ca^{2+} ions. Ca^{2+} ions then diffuse from the sarcoplasmic reticulum into the sarcoplasm.

(c) Ca^{2+} ions released from the sarcoplasmic reticulum bind to troponin in the actin myofilament.

(d) As a result of Ca^{2+} ion binding to troponin, tropomyosin moves deeper into the groove along the actin myofilament and exposes active sites on the actin molecules to which myosin can bind to form cross bridges.

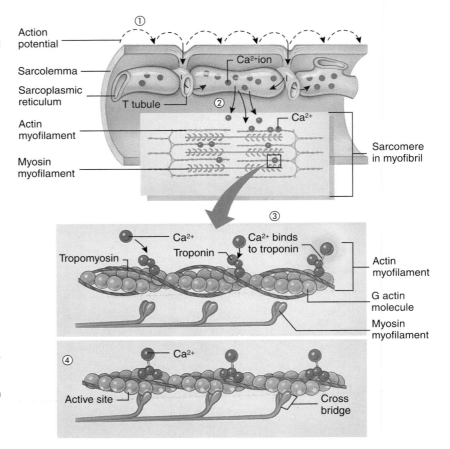

Action Potentials and Muscle Contraction
Figure 10.12

(a) During contraction of a muscle, Ca²⁺ ions bind to troponin, causing exposure of active sites on actin myofilaments.

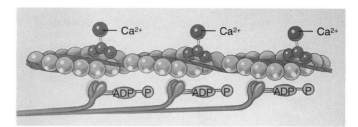

(b) The myosin molecules attach to the exposed active sites on the actin myofilaments, and phosphate is released from the myosin head.

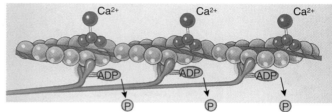

(c) Energy stored in the head of the myosin myofilament is used to move the head of the myosin molecule. Movement of the head, the power stroke, causes the actin myofilament to slide past the myosin myofilament. ADP is released from the myosin head.

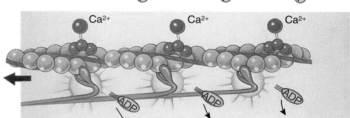

(d) Another ATP molecule binds to the myosin head.

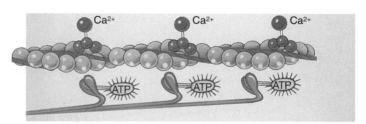

(e) The ATP is broken down to ADP and phosphate, which remain bound to the myosin head, the head of the myosin molecule returns to its resting position, the recovery stroke, and energy is stored in the head of the myosin molecule. If Ca²⁺ ions are still attached to troponin, cross bridge formation and movement are repeated (see **a** through **e**). This cycle occurs many times during a muscle contraction.

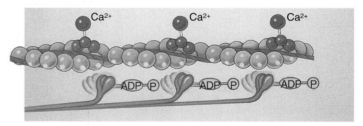

Breakdown of ATP and Cross Bridge Movement
Figure 10.13

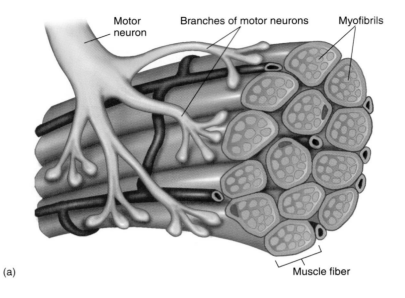

Motor neuron

Branches of motor neurons

Myofibrils

Muscle fiber

(a)

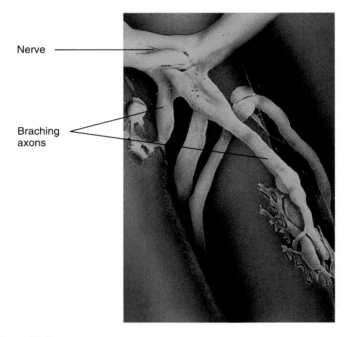

Nerve

Braching axons

(b)

The Motor Unit
Figure 10.15

There is an optimal muscle length at which the muscle produces a maximal tension in response to a maximal stimulus.

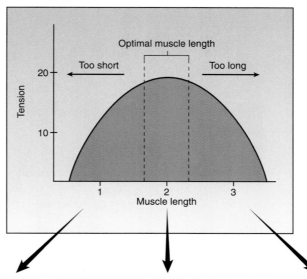

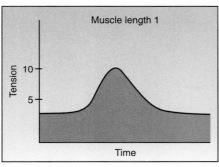

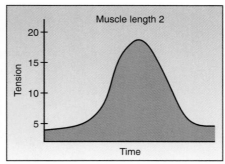

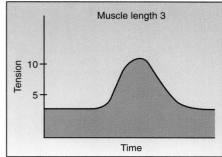

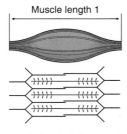

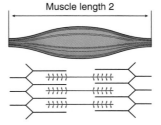

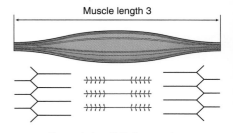

At muscle length 1, the muscle is not stretched, and the tension produced when the muscle contracts is small because the actin and myosin myofilaments are already overlapping nearly as much as they can and the sacromere cannot shorten much more.

At muscle length 2, the muscle is optimally stretched, and the tension produced when the muscle contracts is maximal because the number of cross bridges that can form is maximal.

At muscle length 3, the muscle is stretched severely, and the tension produced is small because the actin and myosin myofilaments only slightly overlap and the number of cross bridges that can form is small.

Muscle Length and Tension
Figure 10.19

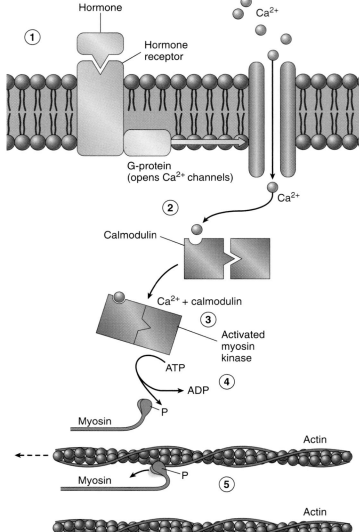

1. Either a hormone combines with a hormone receptor, or depolarization of the cell membrane opens Ca^{2+} ion channels in the plasma membrane.

2. Ca^{2+} ions diffuse through Ca^{2+} ion channels and combine with calmodulin.

3. Calmodulin with a calcium ion bound to it binds with myosin kinase and activates it.

4. Activated myosin kinase attaches phosphate from ATP to myosin heads.

5. A cycle of cross bridge formation, movement, detachment, and cross bridge formation occurs.

6. Relaxation occurs when myosin phosphatase removes phosphate from myosin.

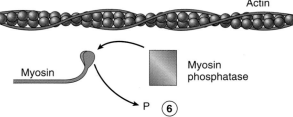

Calcium Ions in Smooth Muscle
Figure 10.22

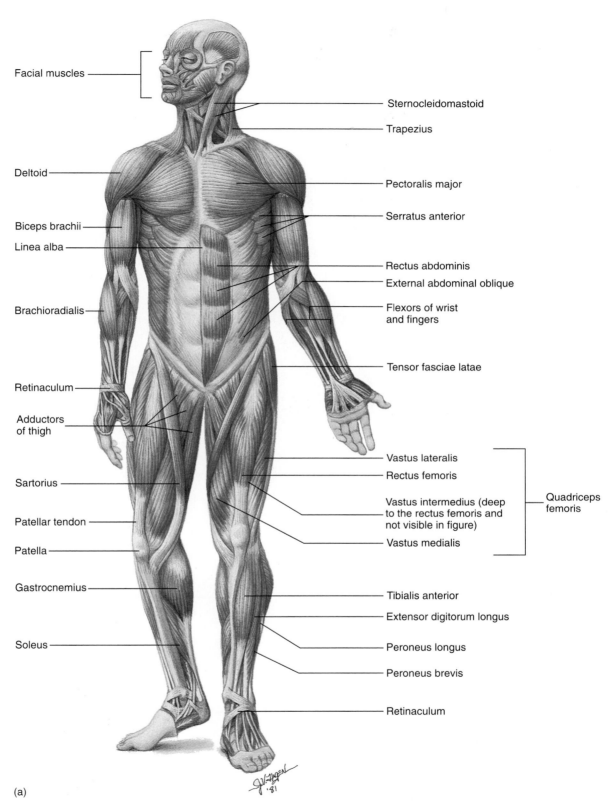

Facial muscles

Sternocleidomastoid

Trapezius

Deltoid

Pectoralis major

Serratus anterior

Biceps brachii

Linea alba

Rectus abdominis

External abdominal oblique

Brachioradialis

Flexors of wrist and fingers

Tensor fasciae latae

Retinaculum

Adductors of thigh

Vastus lateralis

Rectus femoris

Sartorius

Vastus intermedius (deep to the rectus femoris and not visible in figure)

Patellar tendon

Quadriceps femoris

Vastus medialis

Patella

Gastrocnemius

Tibialis anterior

Extensor digitorum longus

Peroneus longus

Soleus

Peroneus brevis

Retinaculum

(a)

General Overview of the Body Musculature (Anterior View)
Figure 11.3a

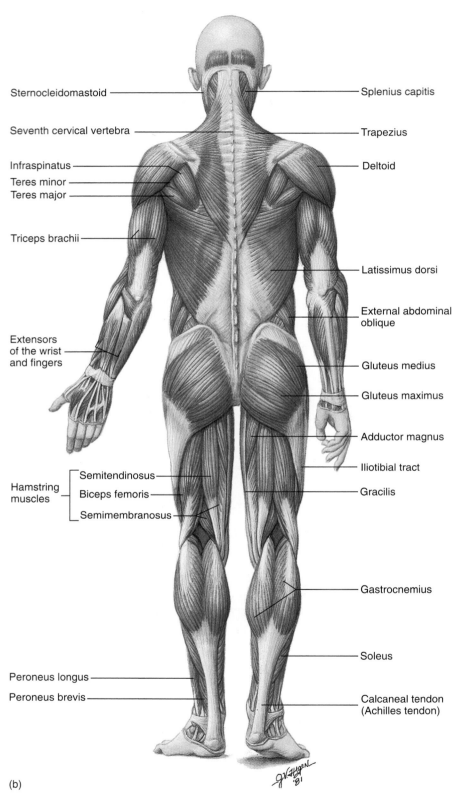

Sternocleidomastoid

Seventh cervical vertebra

Infraspinatus

Teres minor

Teres major

Triceps brachii

Extensors
of the wrist
and fingers

Hamstring
muscles

Semitendinosus

Biceps femoris

Semimembranosus

Peroneus longus

Peroneus brevis

Splenius capitis

Trapezius

Deltoid

Latissimus dorsi

External abdominal
oblique

Gluteus medius

Gluteus maximus

Adductor magnus

Iliotibial tract

Gracilis

Gastrocnemius

Soleus

Calcaneal tendon
(Achilles tendon)

(b)

General Overview of the Body Musculature (Posterior View)
Figure 11.3b

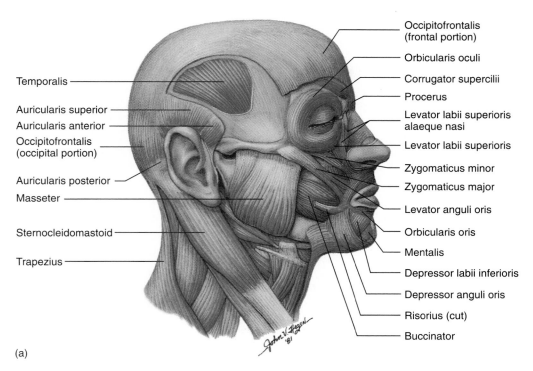

Temporalis

Auricularis superior

Auricularis anterior

Occipitofrontalis
(occipital portion)

Auricularis posterior

Masseter

Sternocleidomastoid

Trapezius

Occipitofrontalis
(frontal portion)

Orbicularis oculi

Corrugator supercilii

Procerus

Levator labii superioris
alaeque nasi

Levator labii superioris

Zygomaticus minor

Zygomaticus major

Levator anguli oris

Orbicularis oris

Mentalis

Depressor labii inferioris

Depressor anguli oris

Risorius (cut)

Buccinator

(a)

Muscles of Facial Expression (Lateral View)
Figure 11.6a

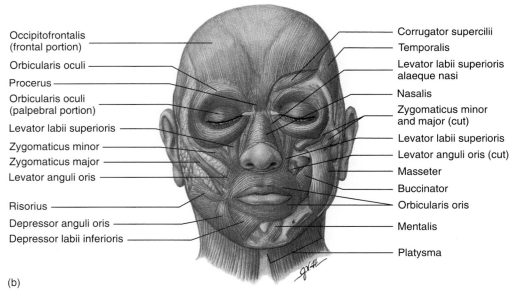

Occipitofrontalis
(frontal portion)

Orbicularis oculi

Procerus

Orbicularis oculi
(palpebral portion)

Levator labii superioris

Zygomaticus minor

Zygomaticus major

Levator anguli oris

Risorius

Depressor anguli oris

Depressor labii inferioris

Corrugator supercilii

Temporalis

Levator labii superioris
alaeque nasi

Nasalis

Zygomaticus minor
and major (cut)

Levator labii superioris

Levator anguli oris (cut)

Masseter

Buccinator

Orbicularis oris

Mentalis

Platysma

(b)

Muscles of Facial Expression (Anterior View)
Figure 11.6b

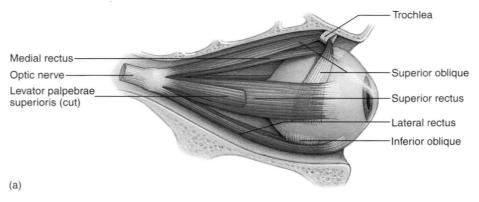

(a)

Muscles Moving the Eyeball (Superior View)
Figure 11.12a

Trochlea

Medial rectus

Optic nerve

Levator palpebrae
superioris (cut)

Superior oblique

Superior rectus

Lateral rectus

Inferior oblique

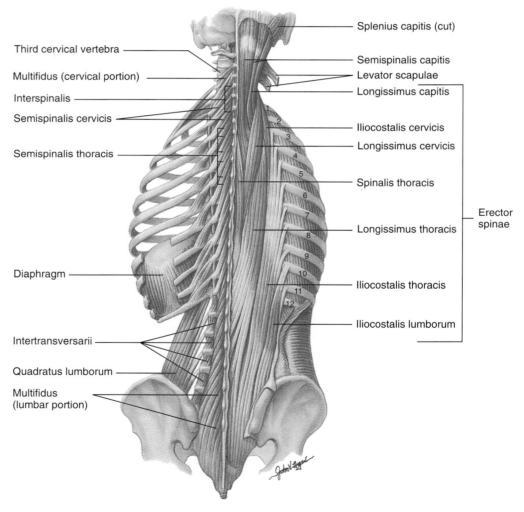

Splenius capitis (cut)

Third cervical vertebra

Multifidus (cervical portion)

Interspinalis

Semispinalis cervicis

Semispinalis thoracis

Semispinalis capitis

Levator scapulae

Longissimus capitis

Iliocostalis cervicis

Longissimus cervicis

Spinalis thoracis

Longissimus thoracis

Iliocostalis thoracis

Iliocostalis lumborum

Erector spinae

Diaphragm

Intertransversarii

Quadratus lumborum

Multifidus
(lumbar portion)

Deep Back Muscles
Figure 11.13

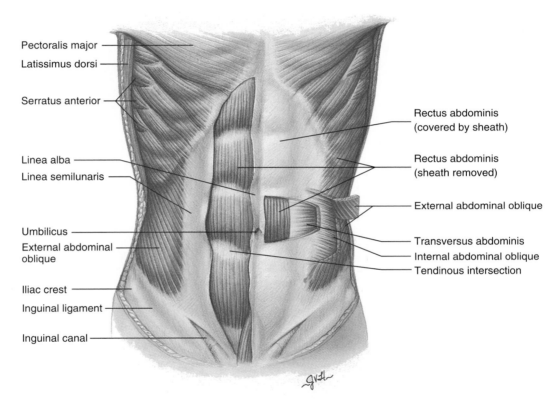

Pectoralis major

Latissimus dorsi

Serratus anterior

Linea alba

Linea semilunaris

Umbilicus

External abdominal
oblique

Iliac crest

Inguinal ligament

Inguinal canal

Rectus abdominis
(covered by sheath)

Rectus abdominis
(sheath removed)

External abdominal oblique

Transversus abdominis

Internal abdominal oblique

Tendinous intersection

Muscles of the Anterior Abdominal Wall
Figure 11.15

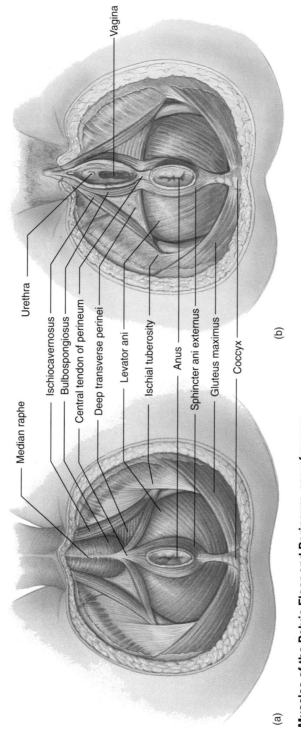

Vagina

Median raphe

Urethra

Ischiocavernosus

Bulbospongiosus

Central tendon of perineum

Deep transverse perinei

Levator ani

Ischial tuberosity

Anus

Sphincter ani externus

Gluteus maximus

Coccyx

(a)

(b)

Muscles of the Pelvic Floor and Perineum, seen from an Inferior View
Figure 11.18

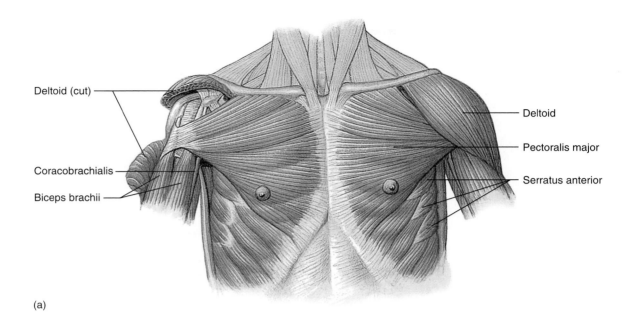

Deltoid (cut)

Coracobrachialis

Biceps brachii

Deltoid

Pectoralis major

Serratus anterior

(a)

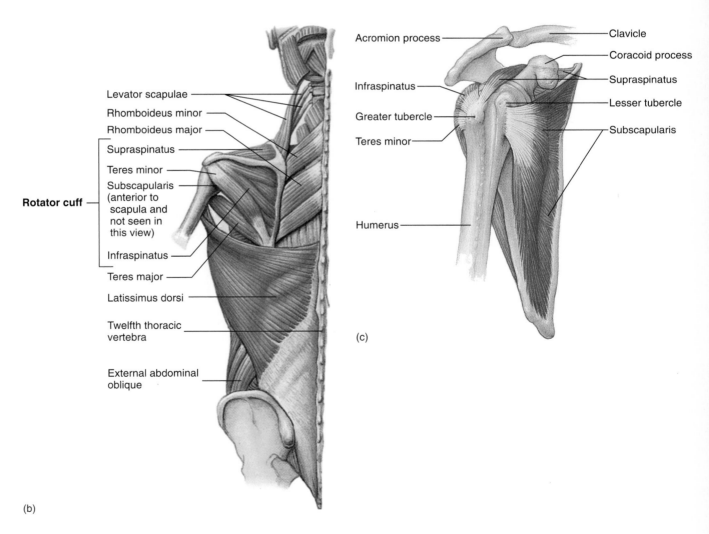

Levator scapulae

Rhomboideus minor

Rhomboideus major

Supraspinatus

Teres minor

Subscapularis
(anterior to
scapula and
not seen in
this view)

Infraspinatus

Teres major

Latissimus dorsi

Twelfth thoracic
vertebra

External abdominal
oblique

Rotator cuff

(b)

Acromion process

Infraspinatus

Greater tubercle

Teres minor

Humerus

Clavicle

Coracoid process

Supraspinatus

Lesser tubercle

Subscapularis

(c)

Muscles Attaching the Upper Limb to the Body
Figure 11.20

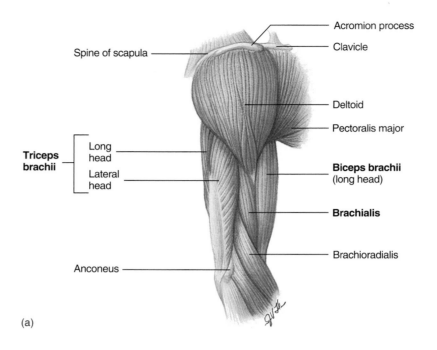

Acromion process

Clavicle

Spine of scapula

Deltoid

Pectoralis major

Triceps brachii

Long head

Lateral head

Biceps brachii (long head)

Brachialis

Brachioradialis

Anconeus

(a)

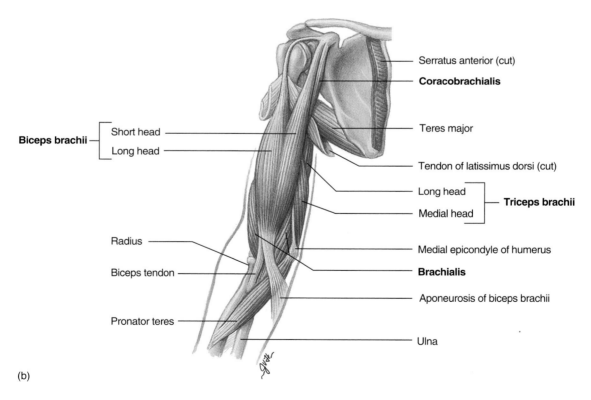

Serratus anterior (cut)

Coracobrachialis

Biceps brachii

Short head

Long head

Teres major

Tendon of latissimus dorsi (cut)

Long head

Medial head

Triceps brachii

Radius

Biceps tendon

Medial epicondyle of humerus

Brachialis

Aponeurosis of biceps brachii

Pronator teres

Ulna

(b)

Muscles of the Arm
Figure 11.22a, b

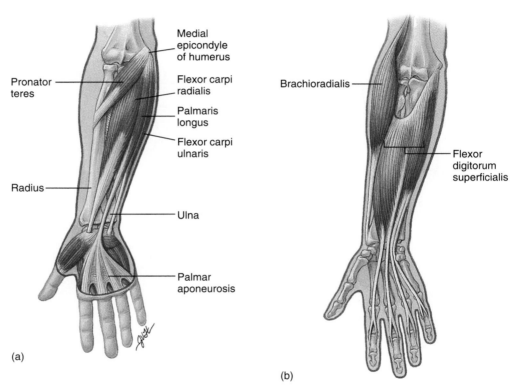

Pronator teres

Medial epicondyle of humerus

Flexor carpi radialis

Palmaris longus

Flexor carpi ulnaris

Radius

Ulna

Palmar aponeurosis

(a)

Brachioradialis

Flexor digitorum superficialis

(b)

Muscles of the Forearm
Figure 11.23a, b

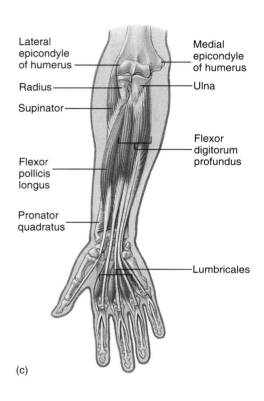

(c)

Lateral epicondyle of humerus

Radius

Supinator

Flexor pollicis longus

Pronator quadratus

Medial epicondyle of humerus

Ulna

Flexor digitorum profundus

Lumbricales

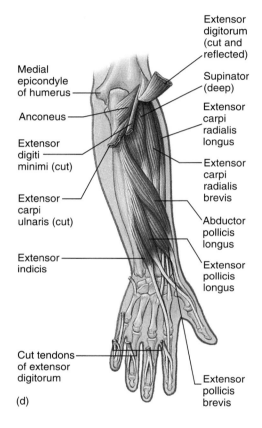

(d)

Medial epicondyle of humerus

Anconeus

Extensor digiti minimi (cut)

Extensor carpi ulnaris (cut)

Extensor indicis

Cut tendons of extensor digitorum

Extensor digitorum (cut and reflected)

Supinator (deep)

Extensor carpi radialis longus

Extensor carpi radialis brevis

Abductor pollicis longus

Extensor pollicis longus

Extensor pollicis brevis

Muscles of the Forearm
Figure 11.23c, d

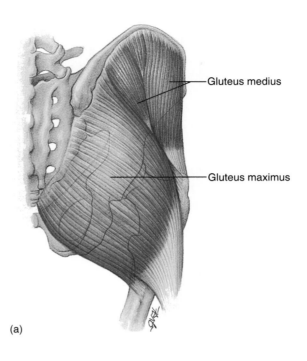

Gluteus medius

Gluteus maximus

(a)

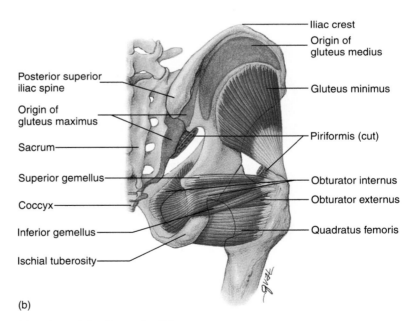

Iliac crest

Origin of
gluteus medius

Posterior superior
iliac spine

Gluteus minimus

Origin of
gluteus maximus

Sacrum

Piriformis (cut)

Superior gemellus

Obturator internus

Coccyx

Obturator externus

Inferior gemellus

Quadratus femoris

Ischial tuberosity

(b)

Muscles of the Posterior Hip
Figure 11.26

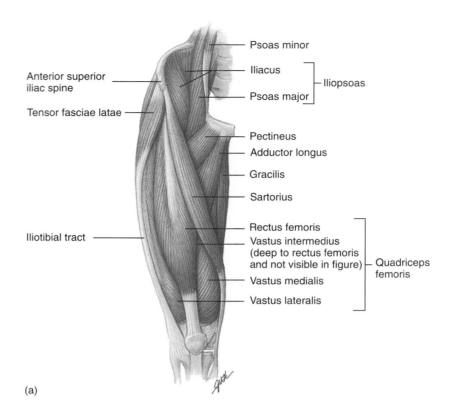

Psoas minor

Anterior superior
iliac spine

Iliacus

Psoas major

Iliopsoas

Tensor fasciae latae

Pectineus

Adductor longus

Gracilis

Sartorius

Rectus femoris

Vastus intermedius
(deep to rectus femoris
and not visible in figure)

Quadriceps
femoris

Iliotibial tract

Vastus medialis

Vastus lateralis

(a)

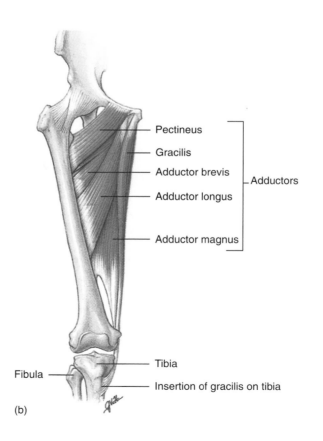

Pectineus

Gracilis

Adductor brevis

Adductors

Adductor longus

Adductor magnus

Tibia

Fibula

Insertion of gracilis on tibia

(b)

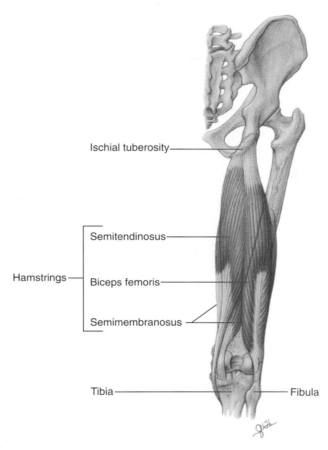

Ischial tuberosity

Semitendinosus

Hamstrings

Biceps femoris

Semimembranosus

Tibia

Fibula

Posterior Muscles of the Right Thigh
Figure 11.28

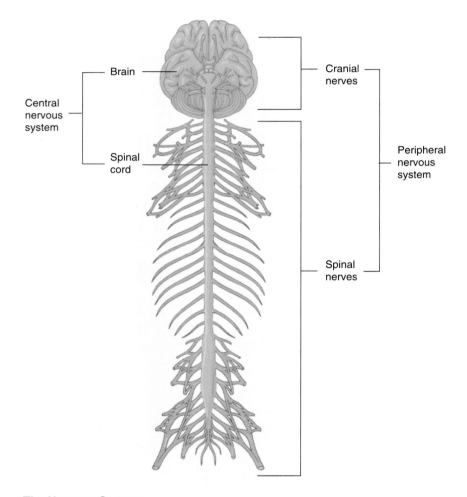

The Nervous System
Figure 12.1

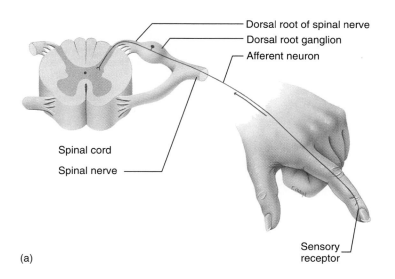

(a)

Dorsal root of spinal nerve
Dorsal root ganglion
Afferent neuron
Spinal cord
Spinal nerve
Sensory receptor

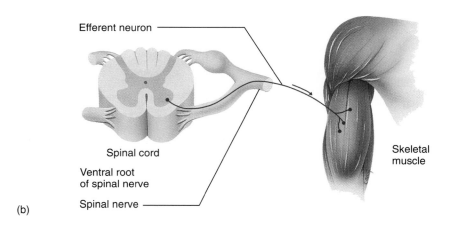

(b)

Efferent neuron
Spinal cord
Ventral root of spinal nerve
Spinal nerve
Skeletal muscle

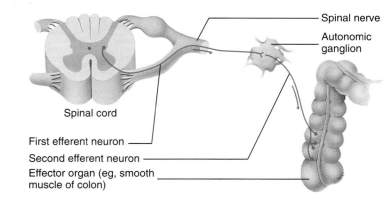

(c)

Spinal nerve
Autonomic ganglion
Spinal cord
First efferent neuron
Second efferent neuron
Effector organ (eg, smooth muscle of colon)

Divisions of the Peripheral Nervous System
Figure 12.3

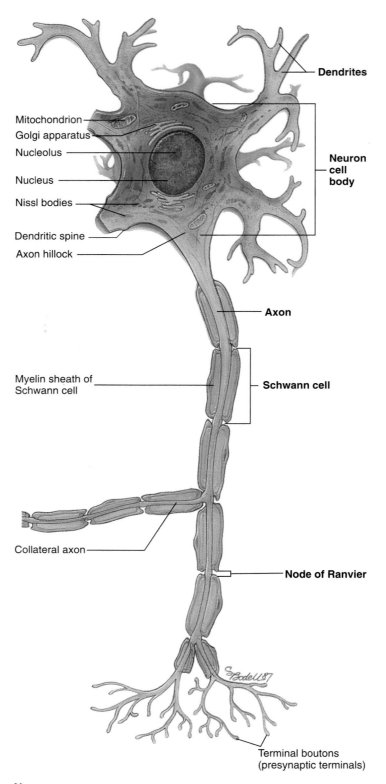

Mitochondrion
Golgi apparatus
Nucleolus
Nucleus
Nissl bodies
Dendritic spine
Axon hillock

Dendrites

Neuron cell body

Axon

Myelin sheath of Schwann cell

Schwann cell

Collateral axon

Node of Ranvier

Terminal boutons (presynaptic terminals)

Neuron
Figure 12.4

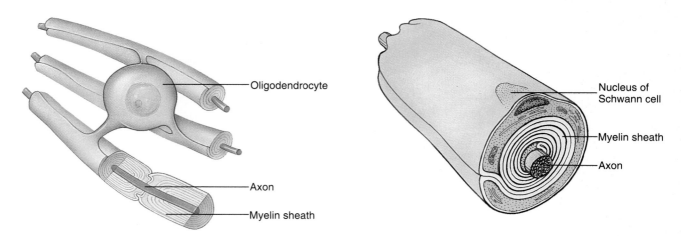

Oligodendrocyte

Axon

Myelin sheath

Nucleus of
Schwann cell

Myelin sheath

Axon

Oligodendrocyte, Neurolemmocyte
Figure 12.9, 12.10

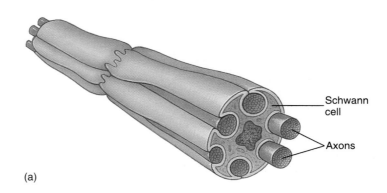

Schwann
cell

Axons

(a)

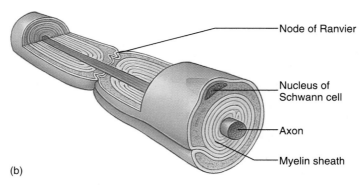

Node of Ranvier

Nucleus of
Schwann cell

Axon

Myelin sheath

(b)

Comparison of Myelinated and Unmyelinated Axons
Figure 12.12

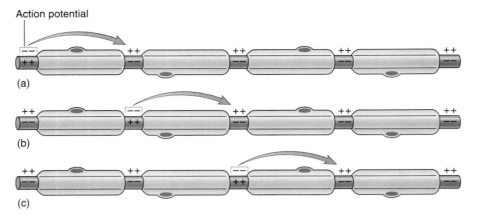

(a)

Action potential

(b)

(c)

Saltatory Conduction
Figure 12.13

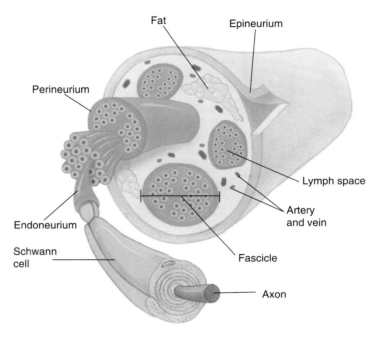

Fat

Epineurium

Perineurium

Lymph space

Endoneurium

Artery
and vein

Schwann
cell

Fascicle

Axon

Nerve
Figure 12.14

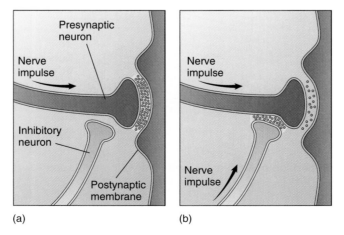

Presynaptic
neuron

Nerve
impulse

Nerve
impulse

Inhibitory
neuron

Postynaptic
membrane

Nerve
impulse

(a)

(b)

Presynaptic Inhibition at an Axo-Axonic Synapse
Figure 12.17

(a) **Spatial summation.** The two local depolarizations produced at 1 and 2 by action potentials that arrive simultaneously summate at the axon hillock to produce a local depolarization that exceeds threshold, resulting in an action potential.

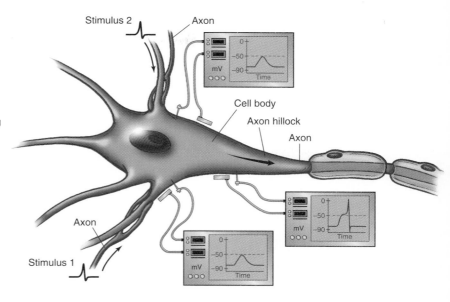

(b) **Temporal summation.** Two action potentials arrive in close succession at the presynaptic membrane. Before the first local depolarization returns to threshold, the second is produced. They summate to exceed threshold and produce an action potential.

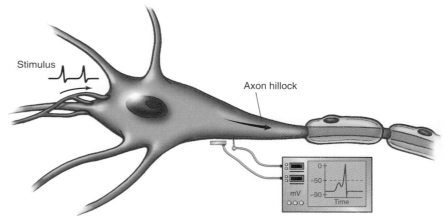

(c) Combined spatial and temporal summation with both excitatory postsynaptic potentials and inhibitory postsynaptic potentials. The outcome, which is the product of summation, is determined by which influence is greater.

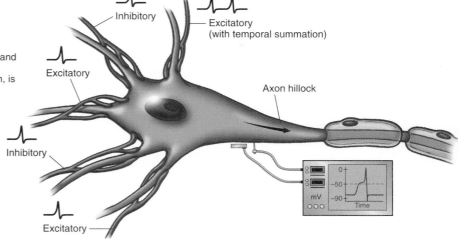

Summation
Figure 12.18

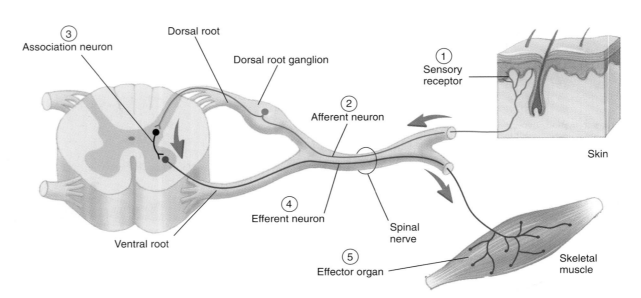

③ Association neuron

Dorsal root

Dorsal root ganglion

① Sensory receptor

② Afferent neuron

Skin

④ Efferent neuron

Spinal nerve

Ventral root

⑤ Effector organ

Skeletal muscle

Reflex Arc
Figure 12.19

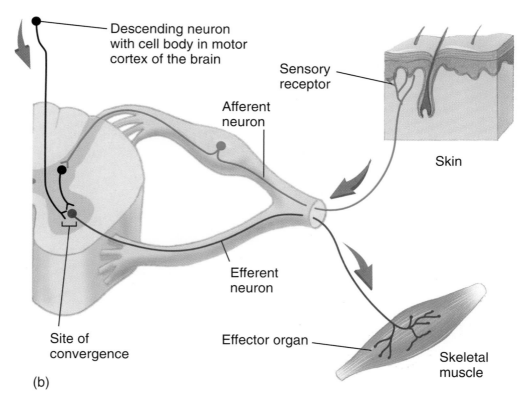

Descending neuron with cell body in motor cortex of the brain

Afferent neuron

Sensory receptor

Skin

Efferent neuron

Site of convergence

Effector organ

Skeletal muscle

(b)

Convergent Pathways
Figure 12.20b

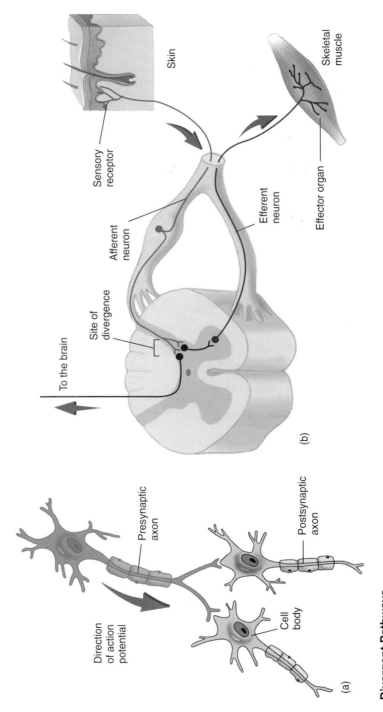

Divergent Pathways
Figure 12.21

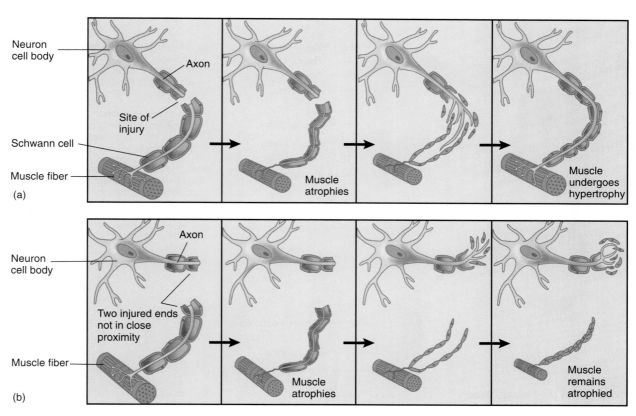

Changes That Occur in an Injured Nerve Fiber
Figure 12A

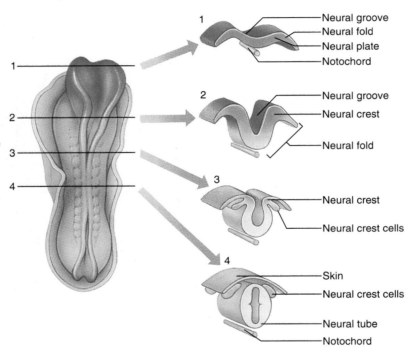

Neural groove
Neural fold
Neural plate
Notochord

Neural groove
Neural crest
Neural fold

Neural crest
Neural crest cells

Skin
Neural crest cells
Neural tube
Notochord

1. The neural plate is a thickened area on the surface of the embryo, overlying the notochord. Neural folds form in the lateral regions of the neural plate and a neural groove forms in the center.

2. The neural folds become more prominent, and the groove deepens. The crest of each fold is referred to as the neural crest.

3. As the neural folds continue to increase in height, neural crest cells begin to migrate away from the under surfaces of the neural crests.

4. The neural crests fuse in the midline to form the neural tube.

Formation of the Neural Tube
Figure 13.2

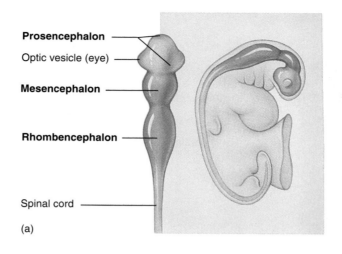

(a)

Labels for (a):
- Prosencephalon
- Optic vesicle (eye)
- Mesencephalon
- Rhombencephalon
- Spinal cord

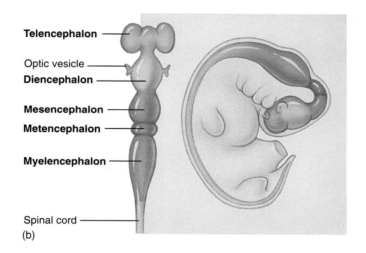

(b)

Labels for (b):
- Telencephalon
- Optic vesicle
- Diencephalon
- Mesencephalon
- Metencephalon
- Myelencephalon
- Spinal cord

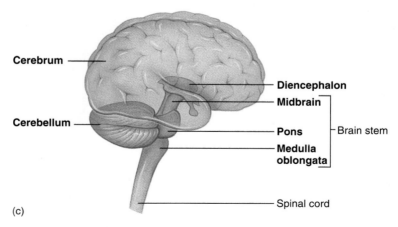

(c)

Labels for (c):
- Cerebrum
- Cerebellum
- Diencephalon
- Midbrain
- Pons
- Medulla oblongata
- Brain stem
- Spinal cord

Development of the Brain Segments and Ventricles
Figure 13.3

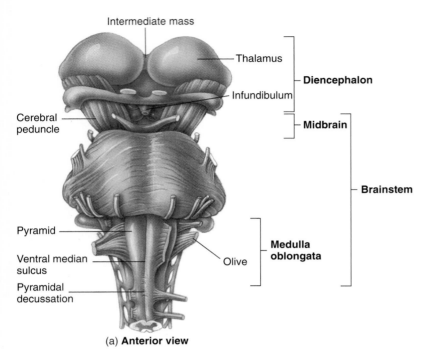

Intermediate mass

Thalamus

Diencephalon

Infundibulum

Cerebral peduncle

Midbrain

Brainstem

Pyramid

Ventral median sulcus

Olive

Medulla oblongata

Pyramidal decussation

(a) **Anterior view**

Brainstem (Anterior View)
Figure 13.5a

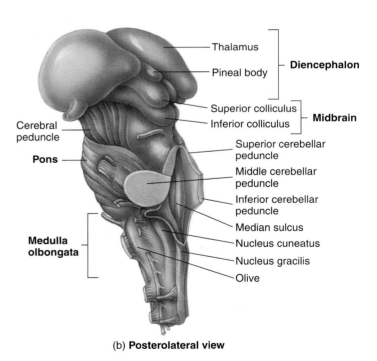

Thalamus

Pineal body

Diencephalon

Superior colliculus

Inferior colliculus

Midbrain

Cerebral peduncle

Superior cerebellar peduncle

Pons

Middle cerebellar peduncle

Inferior cerebellar peduncle

Median sulcus

Medulla olbongata

Nucleus cuneatus

Nucleus gracilis

Olive

(b) **Posterolateral view**

Brainstem (Posterolateral View)
Figure 13.5b

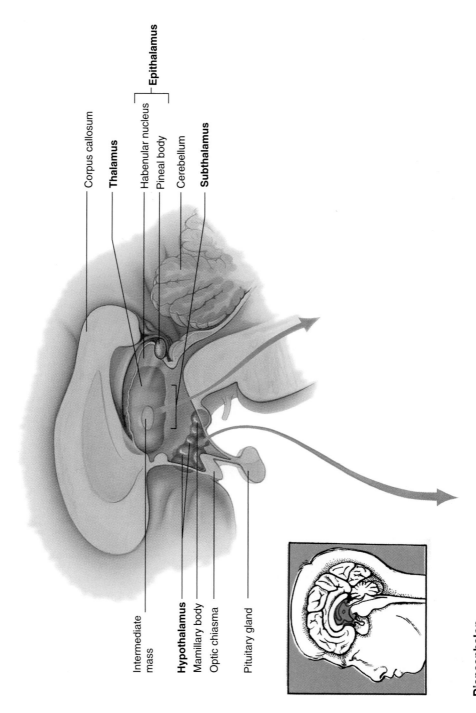

Corpus callosum

Thalamus

Habenular nucleus

Pineal body   **Epithalamus**

Cerebellum

Subthalamus

Intermediate
mass

Hypothalamus

Mamillary body

Optic chiasma

Pituitary gland

Diencephalon
Figure 13.7a

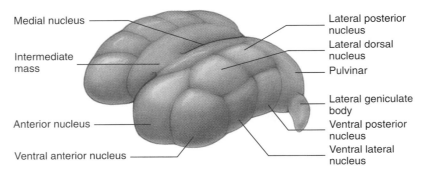

Medial nucleus

Intermediate mass

Anterior nucleus

Ventral anterior nucleus

Lateral posterior nucleus

Lateral dorsal nucleus

Pulvinar

Lateral geniculate body

Ventral posterior nucleus

Ventral lateral nucleus

Diencephalon (Thalamus)
Figure 13.7b

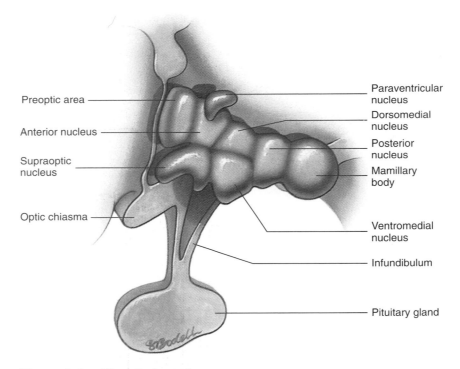

Preoptic area

Anterior nucleus

Supraoptic nucleus

Optic chiasma

Paraventricular nucleus

Dorsomedial nucleus

Posterior nucleus

Mamillary body

Ventromedial nucleus

Infundibulum

Pituitary gland

Diencephalon (Hypothalamus)
Figure 13.7c

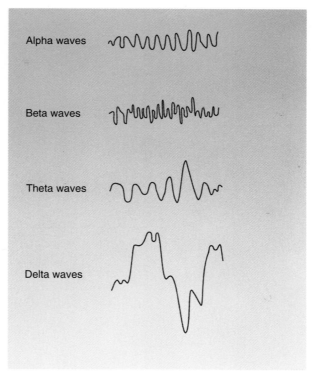

(b)

Tracings from EEGs
Figure 13.13b

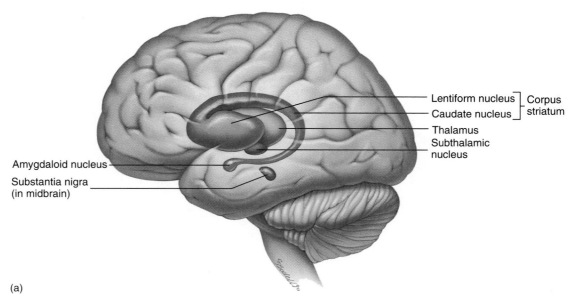

(a)

Basal Ganglia of the Left Hemisphere
Figure 13.14a

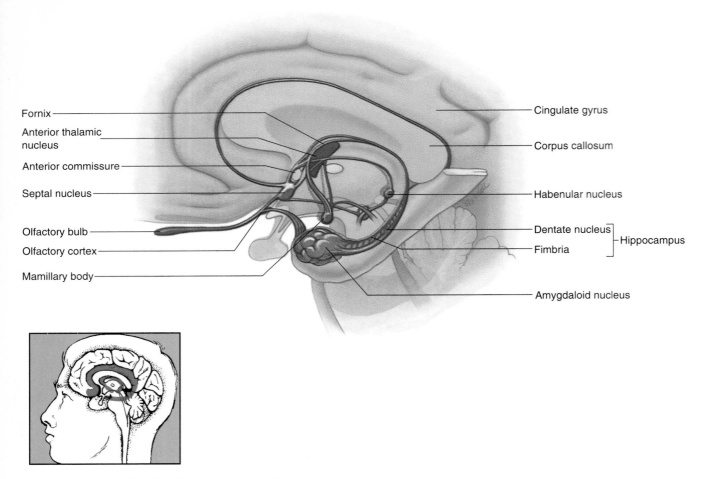

Fornix

Anterior thalamic nucleus

Anterior commissure

Septal nucleus

Olfactory bulb

Olfactory cortex

Mamillary body

Cingulate gyrus

Corpus callosum

Habenular nucleus

Dentate nucleus
Fimbria
} Hippocampus

Amygdaloid nucleus

Limbic System of the Right Hemisphere as Seen in a Midsagittal Section
Figure 13.15

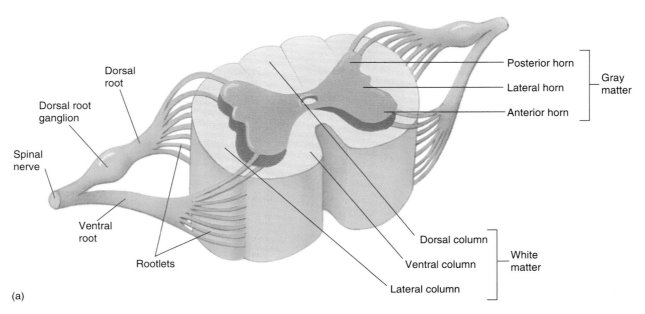

Dorsal root

Dorsal root ganglion

Spinal nerve

Ventral root

Rootlets

Posterior horn

Lateral horn

Anterior horn

} Gray matter

Dorsal column

Ventral column

Lateral column

} White matter

(a)

Cross Section of the Spinal Cord
Figure 13.19

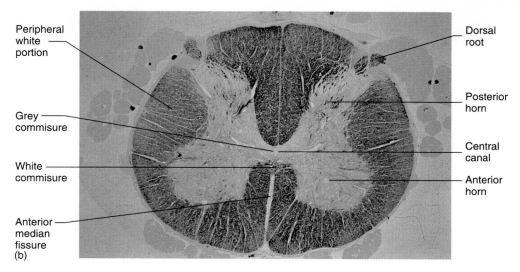

Peripheral white portion

Grey commisure

White commisure

Anterior median fissure
(b)

Dorsal root

Posterior horn

Central canal

Anterior horn

Cross Section of the Spinal Cord
Figure 13.19, *(Continued)*

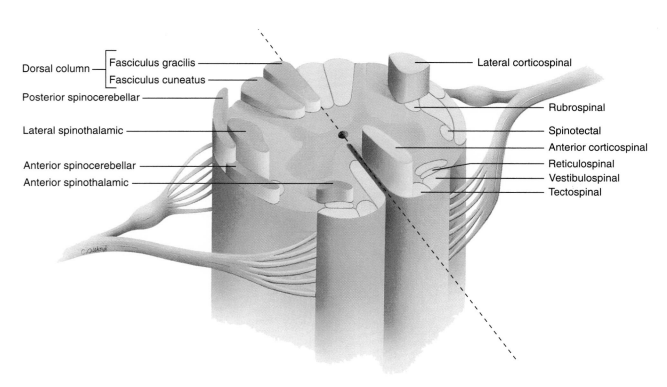

Dorsal column — Fasciculus gracilis
Fasciculus cuneatus

Posterior spinocerebellar

Lateral spinothalamic

Anterior spinocerebellar

Anterior spinothalamic

Lateral corticospinal

Rubrospinal

Spinotectal

Anterior corticospinal

Reticulospinal

Vestibulospinal

Tectospinal

Cross Section of the Spinal Cord Depicting the Pathways
Figure 13.21

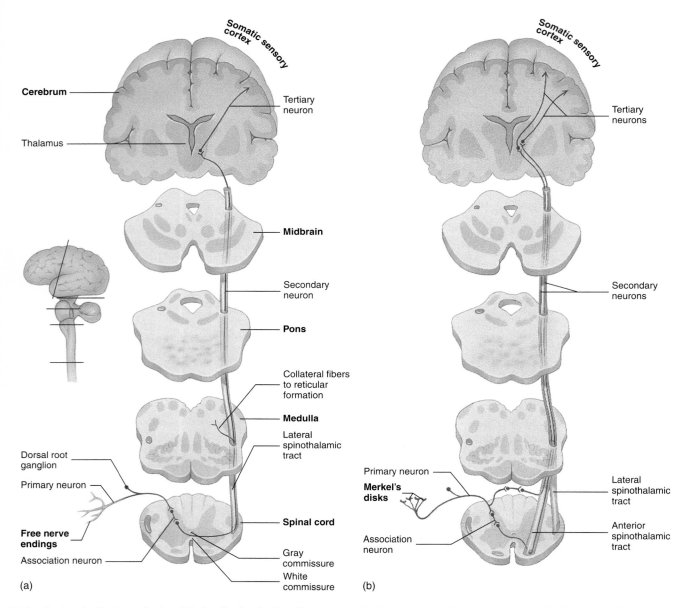

Spinothalamic System (Lateral Spinothalamic Tract)
Figure 13.22a

Spinothalamic System (Anterior Spinothalamic Tract)
Figure 13.22b

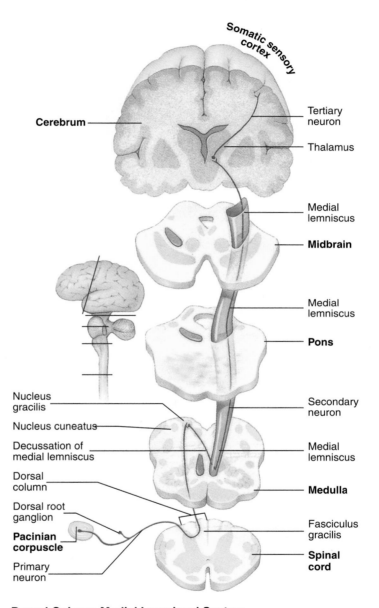

Somatic sensory cortex

Cerebrum

Tertiary neuron

Thalamus

Medial lemniscus

Midbrain

Medial lemniscus

Pons

Nucleus gracilis

Nucleus cuneatus

Decussation of medial lemniscus

Dorsal column

Dorsal root ganglion

Pacinian corpuscle

Primary neuron

Secondary neuron

Medial lemniscus

Medulla

Fasciculus gracilis

Spinal cord

Dorsal Column-Medial Lemniscal System
Figure 13.23

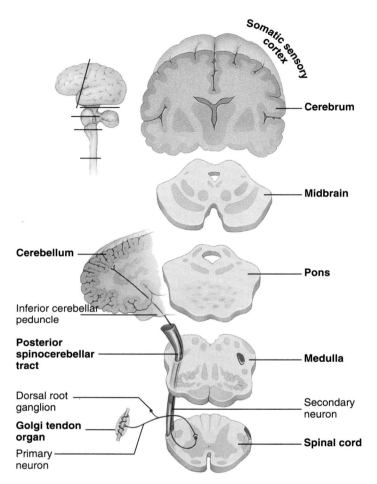

Somatic sensory cortex

—— **Cerebrum**

—— **Midbrain**

Cerebellum ——

—— **Pons**

Inferior cerebellar peduncle

Posterior spinocerebellar tract ——

—— **Medulla**

Dorsal root ganglion ——

Secondary neuron

Golgi tendon organ ——

Primary neuron ——

—— **Spinal cord**

Posterior Spinocerebellar Tract
Figure 13.25

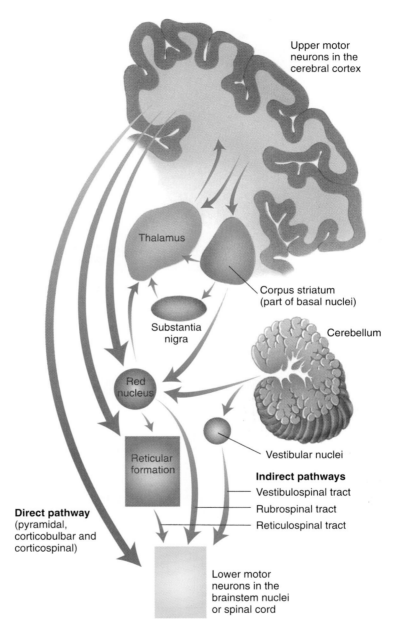

Upper motor
neurons in the
cerebral cortex

Thalamus

Corpus striatum
(part of basal nuclei)

Substantia
nigra

Cerebellum

Red
nucleus

Reticular
formation

Vestibular nuclei

Indirect pathways

Vestibulospinal tract

Rubrospinal tract

Reticulospinal tract

Direct pathway
(pyramidal,
corticobulbar and
corticospinal)

Lower motor
neurons in the
brainstem nuclei
or spinal cord

Descending Pathways
Figure 13.26

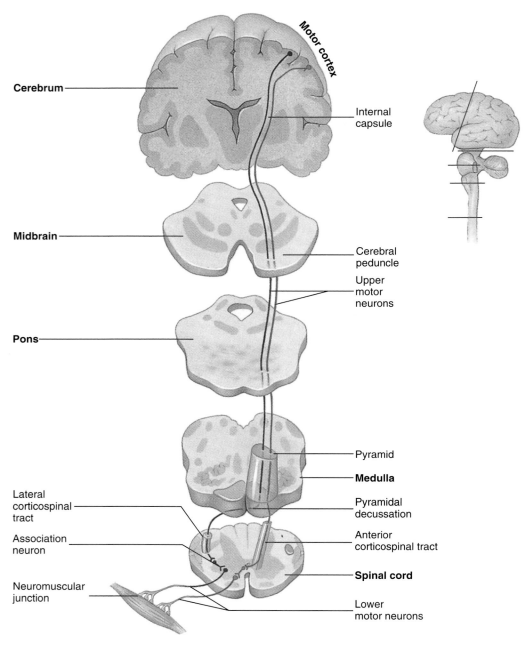

Cerebrum

Motor cortex

Internal capsule

Midbrain

Cerebral peduncle

Upper motor neurons

Pons

Pyramid

Medulla

Lateral corticospinal tract

Pyramidal decussation

Association neuron

Anterior corticospinal tract

Spinal cord

Neuromuscular junction

Lower motor neurons

Direct Pathways
Figure 13.27

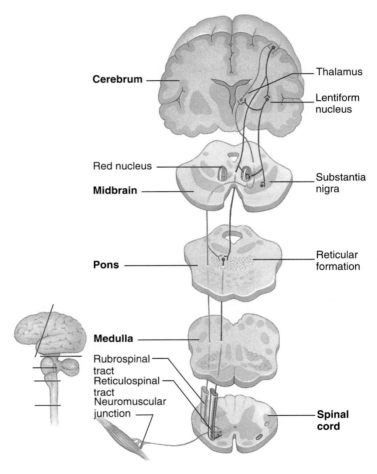

Cerebrum

Thalamus

Lentiform
nucleus

Red nucleus

Midbrain

Substantia
nigra

Pons

Reticular
formation

Medulla

Rubrospinal
tract

Reticulospinal
tract

Neuromuscular
junction

Spinal
cord

Indirect Pathways
Figure 13.28

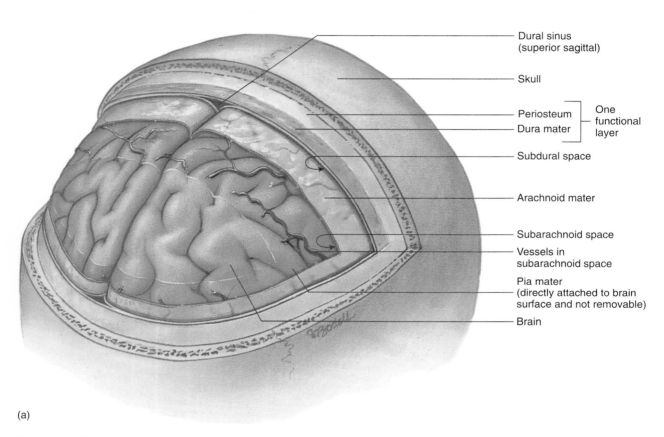

Dural sinus
(superior sagittal)

Skull

Periosteum ⎤ One
          ⎥ functional
Dura mater ⎦ layer

Subdural space

Arachnoid mater

Subarachnoid space

Vessels in
subarachnoid space

Pia mater
(directly attached to brain
surface and not removable)

Brain

(a)

Meninges (Brain)
Figure 13.29a

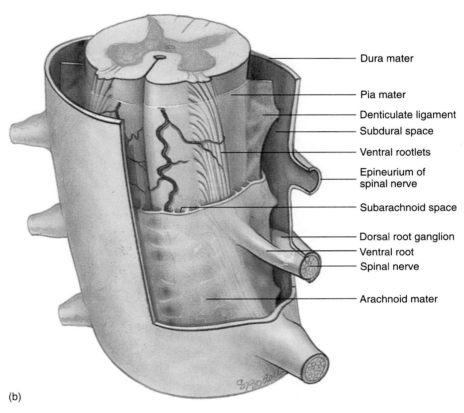

— Dura mater

— Pia mater

— Denticulate ligament

— Subdural space

— Ventral rootlets

Epineurium of
spinal nerve

— Subarachnoid space

— Dorsal root ganglion

— Ventral root

— Spinal nerve

— Arachnoid mater

(b)

Meninges (Spinal Cord)
Figure 13.29b

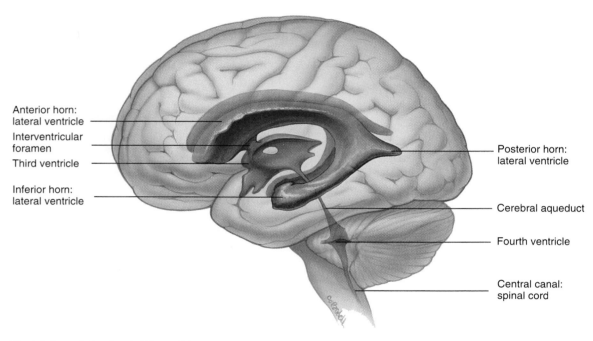

Anterior horn:
lateral ventricle

Interventricular
foramen

Third ventricle

Inferior horn:
lateral ventricle

Posterior horn:
lateral ventricle

Cerebral aqueduct

Fourth ventricle

Central canal:
spinal cord

Ventricles of the Brain Viewed from the Left
Figure 13.30

1. CSF is produced in the choroid plexuses of the ventricles (inset).

2. CSF exits the ventricles through the fourth ventricle and enters the subarachnoid space.

3. CSF returns to the venous circulation through the arachnoid granulations in the superior sagittal sinus (inset).

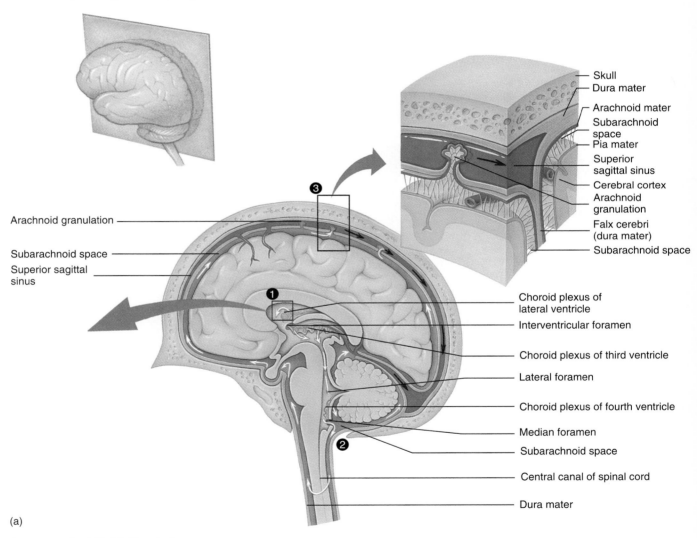

Skull
Dura mater
Arachnoid mater
Subarachnoid space
Pia mater
Superior sagittal sinus
Cerebral cortex
Arachnoid granulation
Falx cerebri (dura mater)
Subarachnoid space

Arachnoid granulation

Subarachnoid space

Superior sagittal sinus

Choroid plexus of lateral ventricle
Interventricular foramen
Choroid plexus of third ventricle
Lateral foramen
Choroid plexus of fourth ventricle
Median foramen
Subarachnoid space
Central canal of spinal cord
Dura mater

(a)

Cerebrospinal Fluid Circulation
Figure 13.31 *(continued)*

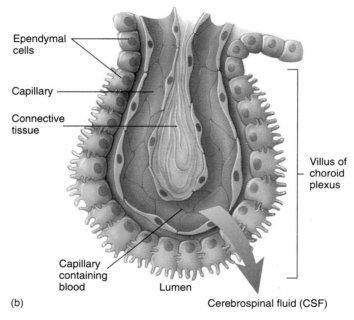

Ependymal cells

Capillary

Connective tissue

Capillary containing blood

Lumen

Villus of choroid plexus

Cerebrospinal fluid (CSF)

(b)

Figure 13.31 *(continued)*

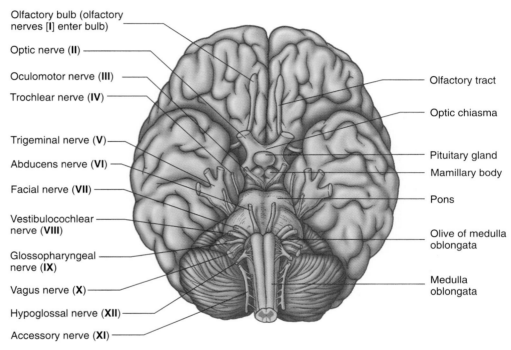

Olfactory bulb (olfactory nerves [**I**] enter bulb)

Optic nerve (**II**)

Oculomotor nerve (**III**)

Trochlear nerve (**IV**)

Trigeminal nerve (**V**)

Abducens nerve (**VI**)

Facial nerve (**VII**)

Vestibulocochlear nerve (**VIII**)

Glossopharyngeal nerve (**IX**)

Vagus nerve (**X**)

Hypoglossal nerve (**XII**)

Accessory nerve (**XI**)

Olfactory tract

Optic chiasma

Pituitary gland

Mamillary body

Pons

Olive of medulla oblongata

Medulla oblongata

Inferior Surface of the Brain Showing the Origin of the Cranial Nerves
Figure 14.1

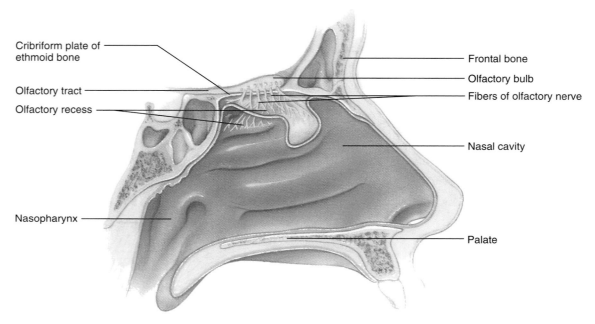

Cribriform plate of ethmoid bone

Frontal bone

Olfactory bulb

Olfactory tract

Fibers of olfactory nerve

Olfactory recess

Nasal cavity

Nasopharynx

Palate

Olfactory Recess and Bulb
Figure 15.3

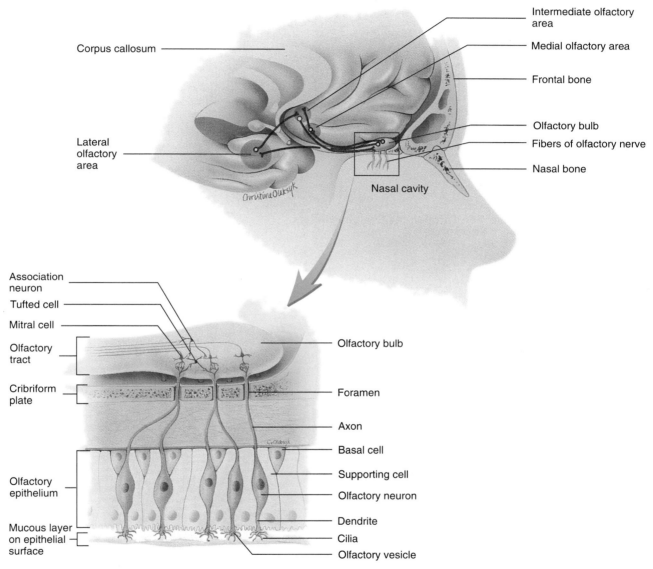

Corpus callosum

Intermediate olfactory area

Medial olfactory area

Frontal bone

Olfactory bulb

Fibers of olfactory nerve

Nasal bone

Lateral olfactory area

Nasal cavity

Association neuron

Tufted cell

Mitral cell

Olfactory tract

Olfactory bulb

Cribriform plate

Foramen

Axon

Basal cell

Supporting cell

Olfactory epithelium

Olfactory neuron

Dendrite

Mucous layer on epithelial surface

Cilia

Olfactory vesicle

Olfactory Epithelium and Olfactory Bulb
Figure 15.4

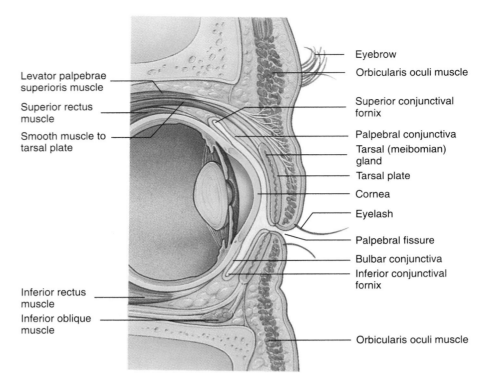

Levator palpebrae superioris muscle

Superior rectus muscle

Smooth muscle to tarsal plate

Inferior rectus muscle

Inferior oblique muscle

Eyebrow

Orbicularis oculi muscle

Superior conjunctival fornix

Palpebral conjunctiva

Tarsal (meibomian) gland

Tarsal plate

Cornea

Eyelash

Palpebral fissure

Bulbar conjunctiva

Inferior conjunctival fornix

Orbicularis oculi muscle

Sagittal Section Through the Eye Showing its Accessory Structures
Figure 15.9

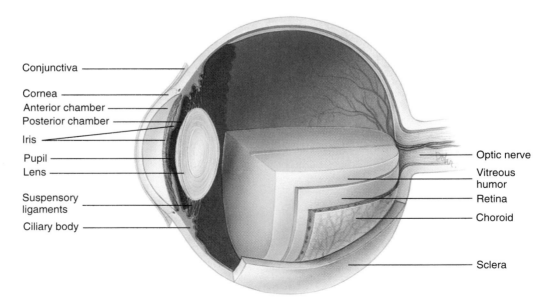

Conjunctiva

Cornea

Anterior chamber

Posterior chamber

Iris

Pupil

Lens

Suspensory ligaments

Ciliary body

Optic nerve

Vitreous humor

Retina

Choroid

Sclera

Sagittal Section of the Eye Demonstrating its Layers
Figure 15.13

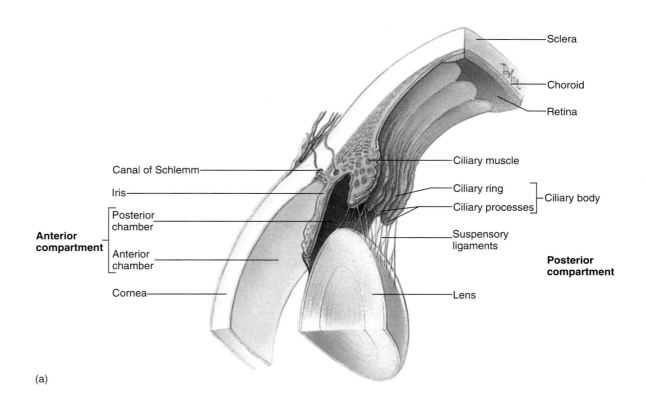

(a)

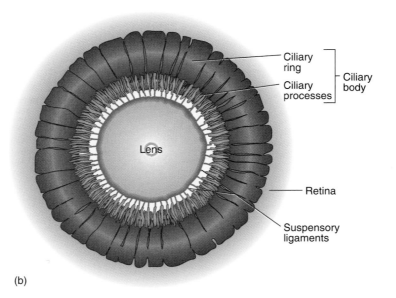

(b)

Lens, Cornea, Iris, and Ciliary Body
Figure 15.14

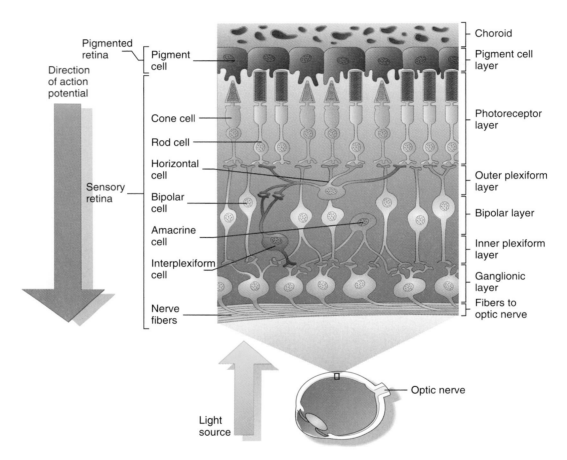

Pigmented retina

Pigment cell

Direction of action potential

Cone cell

Rod cell

Horizontal cell

Sensory retina

Bipolar cell

Amacrine cell

Interplexiform cell

Nerve fibers

Choroid

Pigment cell layer

Photoreceptor layer

Outer plexiform layer

Bipolar layer

Inner plexiform layer

Ganglionic layer

Fibers to optic nerve

Light source

Optic nerve

Retina
Figure 15.18

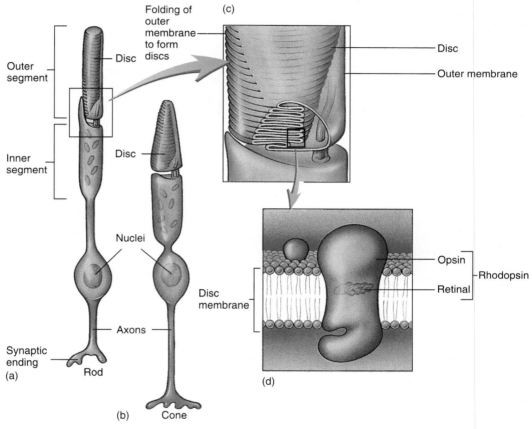

Sensory Receptor Cells of the Retina
Figure 15.19

1. Axons from neurons in the eyes travel through the optic nerves.

2. In the optic chiasma, some of the fibers cross (those from the medial half of the retina) and some do not (those from the lateral half of the retina).

3. The optic tract consists of axons that have passed through the optic chiasma.

4. The axons synapse in the lateral geniculate nuclei of the thalamus. Collateral axons synapse in the superior colliculi.

5. Thalamic neurons form the optic radiations, which project to the visual cortex.

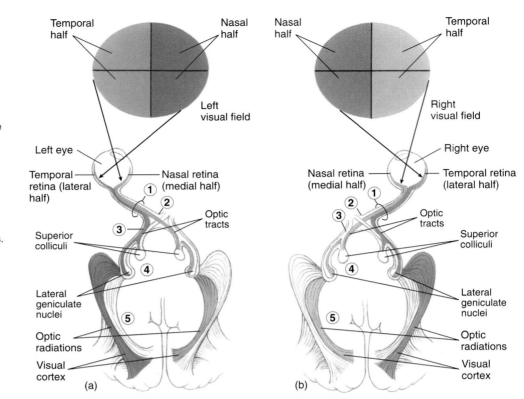

Temporal half · Nasal half · Left visual field · Left eye · Temporal retina (lateral half) · Nasal retina (medial half) · Optic tracts · Superior colliculi · Lateral geniculate nuclei · Optic radiations · Visual cortex · (a)

Nasal half · Temporal half · Right visual field · Right eye · Nasal retina (medial half) · Temporal retina (lateral half) · Optic tracts · Superior colliculi · Lateral geniculate nuclei · Optic radiations · Visual cortex · (b)

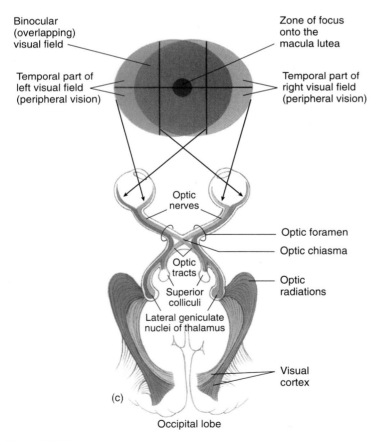

Binocular (overlapping) visual field · Zone of focus onto the macula lutea · Temporal part of left visual field (peripheral vision) · Temporal part of right visual field (peripheral vision) · Optic nerves · Optic foramen · Optic chiasma · Optic tracts · Optic radiations · Superior colliculi · Lateral geniculate nuclei of thalamus · Visual cortex · (c) · Occipital lobe

Visual Pathways
Figure 15.22a-c

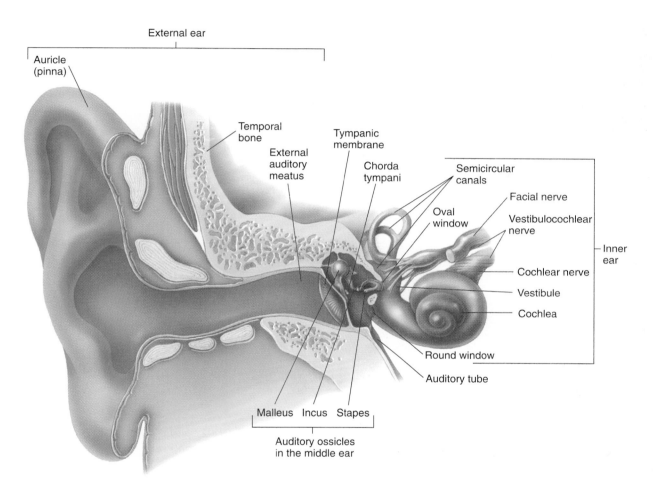

External ear

Auricle
(pinna)

Temporal
bone

External
auditory
meatus

Tympanic
membrane

Chorda
tympani

Semicircular
canals

Oval
window

Facial nerve

Vestibulocochlear
nerve

Cochlear nerve

Vestibule

Cochlea

Round window

Auditory tube

Inner
ear

Malleus   Incus   Stapes

Auditory ossicles
in the middle ear

External, Middle, and Inner Ear
Figure 15.23

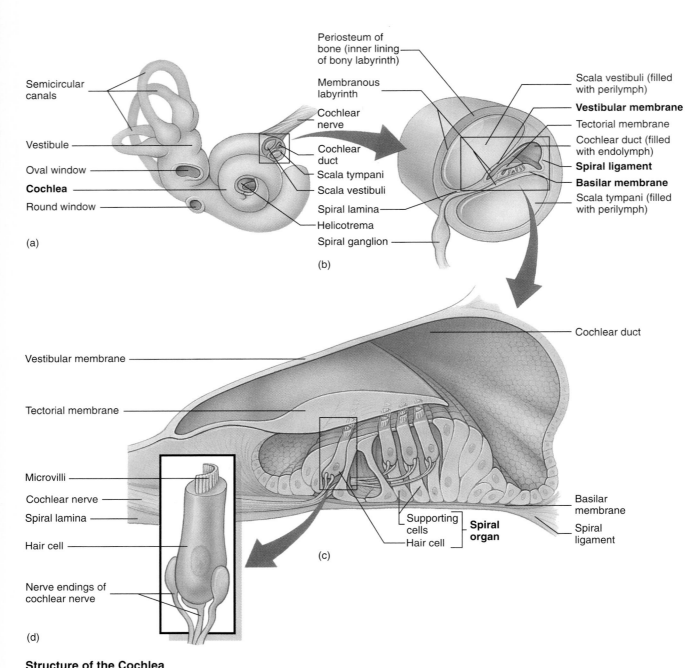

Semicircular canals

Vestibule

Oval window

Cochlea

Round window

(a)

Periosteum of bone (inner lining of bony labyrinth)

Membranous labyrinth

Cochlear nerve

Cochlear duct

Scala tympani

Scala vestibuli

Spiral lamina

Helicotrema

Spiral ganglion

(b)

Scala vestibuli (filled with perilymph)

Vestibular membrane

Tectorial membrane

Cochlear duct (filled with endolymph)

Spiral ligament

Basilar membrane

Scala tympani (filled with perilymph)

Cochlear duct

Vestibular membrane

Tectorial membrane

Microvilli

Cochlear nerve

Spiral lamina

Hair cell

Nerve endings of cochlear nerve

(d)

Supporting cells

Hair cell

Spiral organ

(c)

Basilar membrane

Spiral ligament

Structure of the Cochlea
Figure 15.26

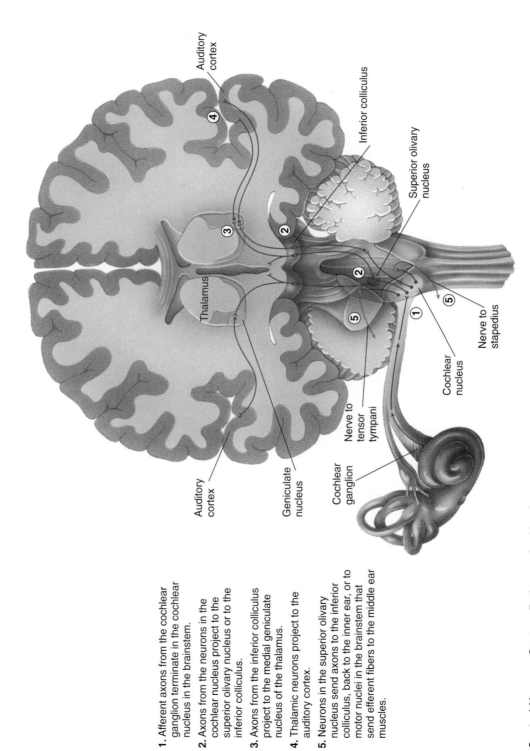

1. Afferent axons from the cochlear ganglion terminate in the cochlear nucleus in the brainstem.

2. Axons from the neurons in the cochlear nucleus project to the superior olivary nucleus or to the inferior colliculus.

3. Axons from the inferior colliculus project to the medial geniculate nucleus of the thalamus.

4. Thalamic neurons project to the auditory cortex.

5. Neurons in the superior olivary nucleus send axons to the inferior colliculus, back to the inner ear, or to motor nuclei in the brainstem that send efferent fibers to the middle ear muscles.

Central Nervous System Pathways for Hearing
Figure 15.32

Auditory cortex

Inferior colliculus

Superior olivary nucleus

Thalamus

Auditory cortex

Geniculate nucleus

Nerve to stapedius

Cochlear nucleus

Nerve to tensor tympani

Cochlear ganglion

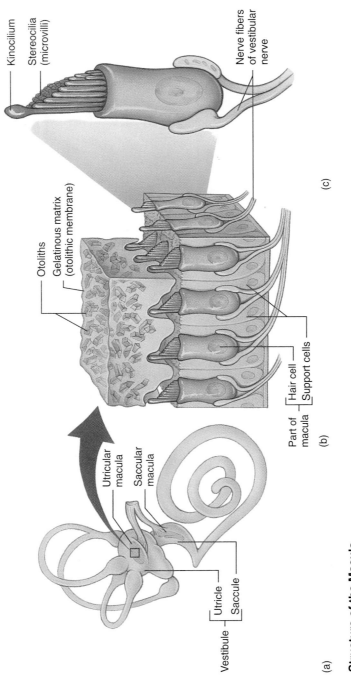

Kinocilium

Stereocilia (microvilli)

Nerve fibers of vestibular nerve

Otoliths

Gelatinous matrix (otolithic membrane)

(c)

Part of macula

Hair cell

Support cells

(b)

Utricular macula

Saccular macula

Vestibule

Utricle

Saccule

(a)

Structure of the Macula
Figure 15.33

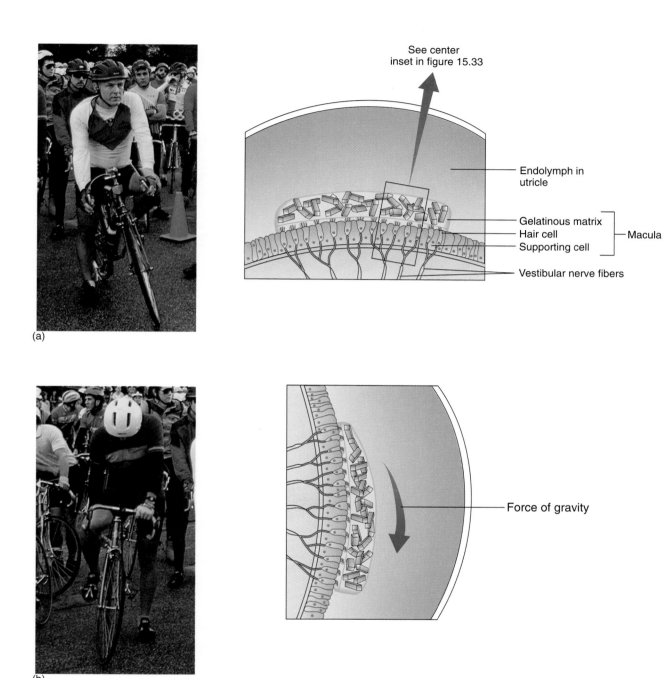

(a)

(b)

See center
inset in figure 15.33

Endolymph in
utricle

Gelatinous matrix
Hair cell — Macula
Supporting cell

Vestibular nerve fibers

Force of gravity

Function of the Vestibule in Maintaining Balance
Figure 15.34

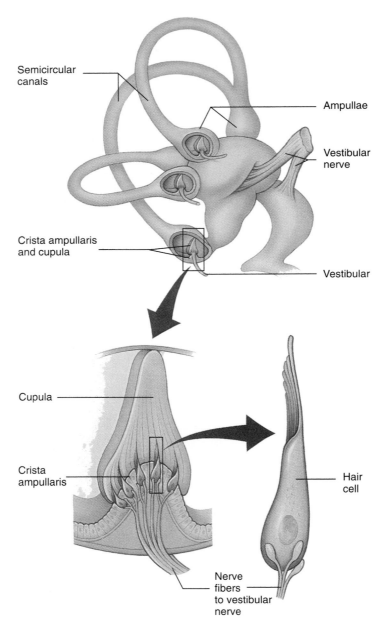

Semicircular canals

Ampullae

Vestibular nerve

Crista ampullaris and cupula

Vestibular

Cupula

Crista ampullaris

Hair cell

Nerve fibers to vestibular nerve

Semicircular Canals
Figure 15.35

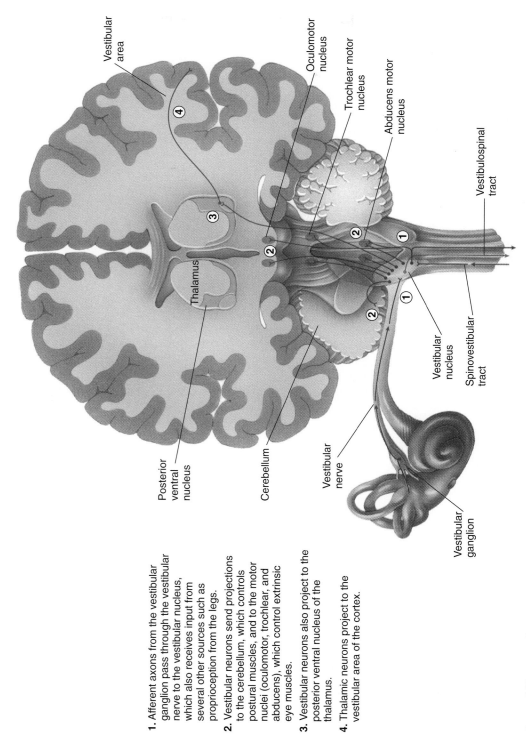

1. Afferent axons from the vestibular ganglion pass through the vestibular nerve to the vestibular nucleus, which also receives input from several other sources such as proprioception from the legs.

2. Vestibular neurons send projections to the cerebellum, which controls postural muscles, and to the motor nuclei (oculomotor, trochlear, and abducens), which control extrinsic eye muscles.

3. Vestibular neurons also project to the posterior ventral nucleus of the thalamus.

4. Thalamic neurons project to the vestibular area of the cortex.

Central Nervous System Pathways for Balance
Figure 15.37

Vestibular area

Oculomotor nucleus

Trochlear motor nucleus

Abducens motor nucleus

Vestibulospinal tract

Posterior ventral nucleus

Thalamus

Cerebellum

Vestibular nucleus

Spinovestibular tract

Vestibular nerve

Vestibular ganglion

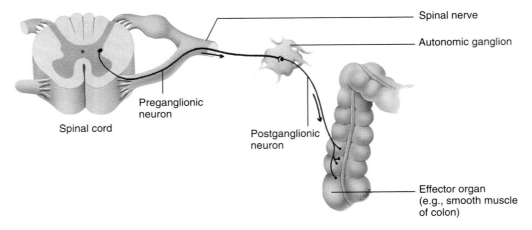

Organization of the Autonomic Nervous System Neurons
Figure 16.1

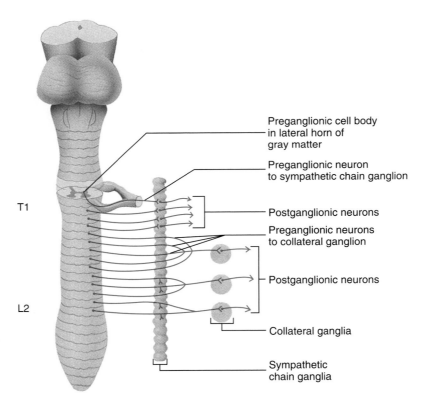

Sympathetic Nervous System
Figure 16.2

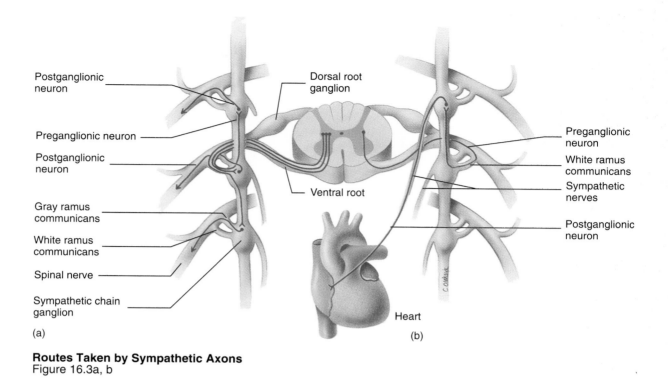

Postganglionic neuron

Preganglionic neuron

Postganglionic neuron

Gray ramus communicans

White ramus communicans

Spinal nerve

Sympathetic chain ganglion

(a)

Dorsal root ganglion

Ventral root

Heart

(b)

Preganglionic neuron

White ramus communicans

Sympathetic nerves

Postganglionic neuron

Routes Taken by Sympathetic Axons
Figure 16.3a, b

Gray ramus communicans

White ramus communicans

Splanchnic nerve

Preganglionic neuron

Collateral ganglia

Postganglionic neuron

(c)

Adrenal gland

Viscera

(d)

White ramus communicans

Preganglionic neuron

Sympathetic chain ganglion

Routes Taken by Sympathetic Axons
Figure 16.3c, d

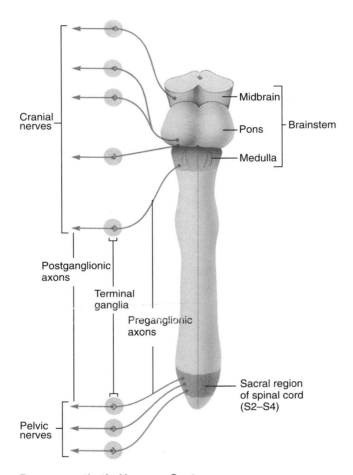

Cranial nerves

Midbrain

Pons — Brainstem

Medulla

Postganglionic axons

Terminal ganglia

Preganglionic axons

Sacral region of spinal cord (S2–S4)

Pelvic nerves

Parasympathetic Nervous System
Figure 16.4

Sympathetic division

Most target tissues innervated by the sympathetic division have adrenergic receptors. When norepinephrine binds to adrenergic receptors, some target tissues are stimulated, and others are inhibited. For example, blood vessels are stimulated to constrict, and stomach glands are inhibited.

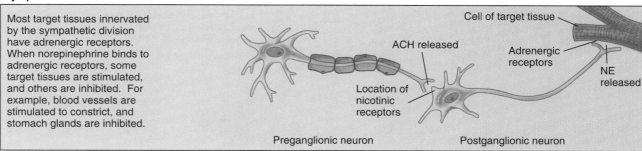

Sympathetic division

Some sympathetic target tissues, such as sweat glands, have muscarinic receptors. Stimulation of sweat glands results in increased sweat production.

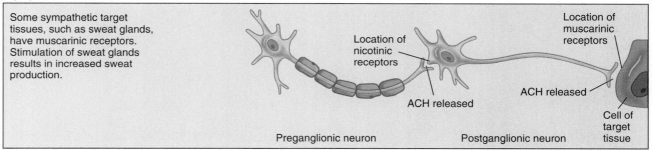

Parasympathetic division

All parasympathetic target tissues have muscarinic receptors. The general response is excitatory, but some target tissues, such as the heart, are inhibited.

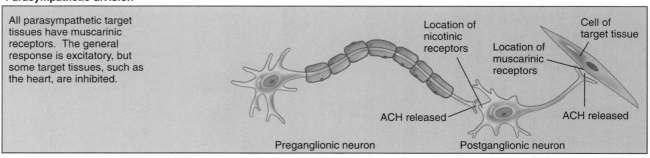

Location of ANS Receptors
Figure 16.5

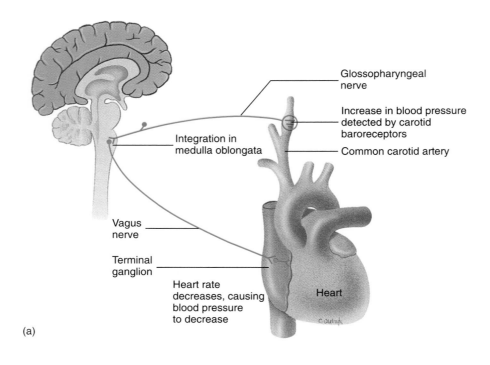

Glossopharyngeal
nerve

Increase in blood pressure
detected by carotid
baroreceptors

Common carotid artery

Integration in
medulla oblongata

Vagus
nerve

Terminal
ganglion

Heart rate
decreases, causing
blood pressure
to decrease

Heart

(a)

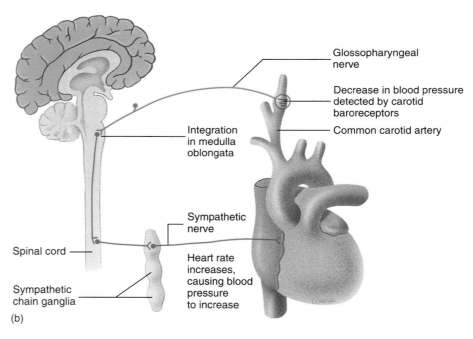

Glossopharyngeal
nerve

Decrease in blood pressure
detected by carotid
baroreceptors

Common carotid artery

Integration
in medulla
oblongata

Sympathetic
nerve

Spinal cord

Heart rate
increases,
causing blood
pressure
to increase

Sympathetic
chain ganglia

(b)

Autonomic Reflexes
Figure 16.6

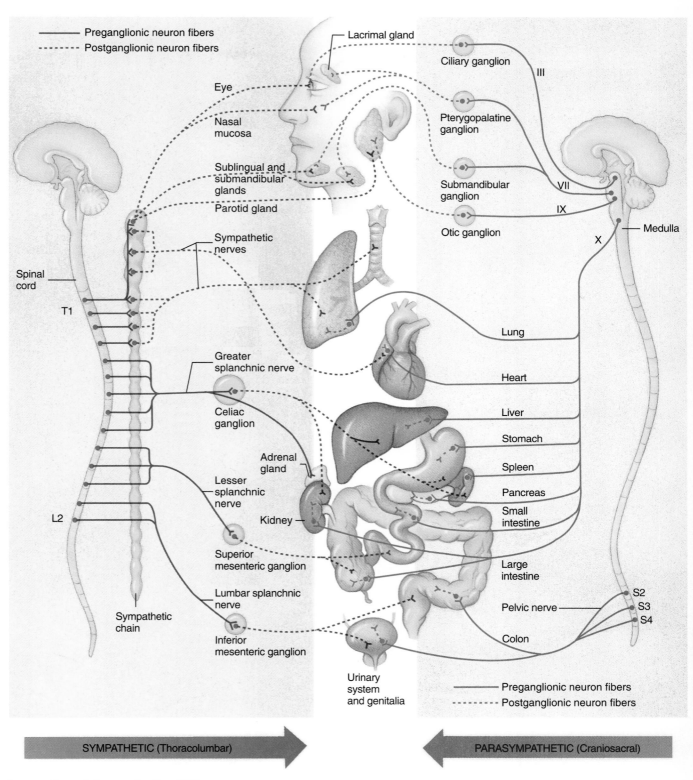

Innervation of Organs by the ANS
Figure 16.8

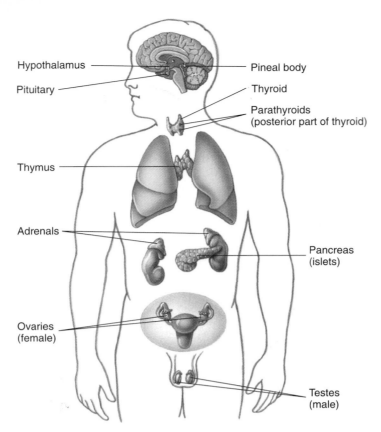

Hypothalamus

Pituitary

Pineal body

Thyroid

Parathyroids
(posterior part of thyroid)

Thymus

Adrenals

Pancreas
(islets)

Ovaries
(female)

Testes
(male)

Location of Major Endocrine Glands
Figure 17.1

1. Stimuli such as stress or exercise activate the sympathetic division of the autonomic nervous system.

2. Sympathetic neurons stimulate the release of epinephrine and smaller amounts of norepinephrine from the adrenal medulla. Epinephrine and norepinephrine prepare the body to respond to stressful conditions.

Once the stressful stimuli are removed, less epinephrine is released as a result of decreased stimulation from the autonomic nervous system.

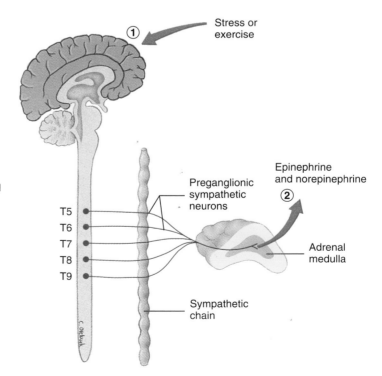

Stress or
exercise

Preganglionic
sympathetic
neurons

Epinephrine
and norepinephrine

Adrenal
medulla

T5
T6
T7
T8
T9

Sympathetic
chain

Sympathetic Nervous System Stimulates the Adrenal Gland
Figure 17.5

1. Thyroid-releasing hormone (TRH) is released from neurons in the hypothalamus and travels to the anterior pituitary gland.

2. TRH stimulates the release of thyroid stimulating hormone (TSH) from the anterior pituitary gland. TSH travels to the thyroid gland.

3. TSH stimulates the secretion of thyroid hormones from the thyroid gland (green arrows).

4. Thyroid hormones act on tissues to produce the usual response to thyroid hormones.

5. Thyroid hormones also act on the hypothalamus and the anterior pituitary to inhibit both TRH secretion and TSH secretion (red arrows).

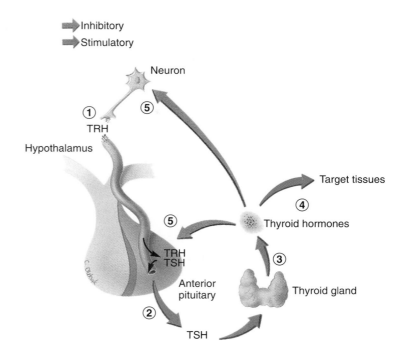

Hormones Can Stimulate or Inhibit the Secretion of Other Hormones
Figure 17.6

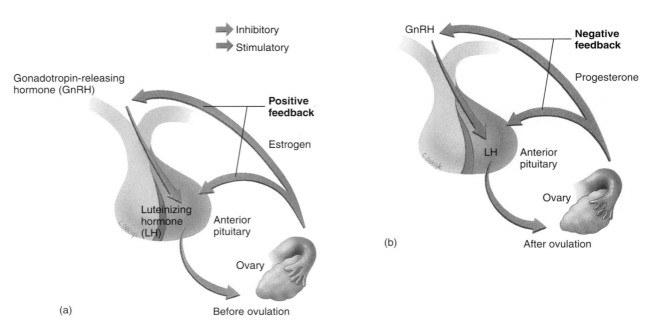

Positive and Negative Feedback
Figure 17.7

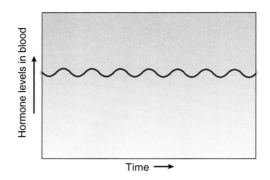

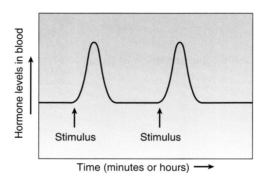

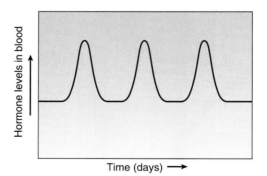

Patterns of Hormone Secretion
Figure 17.8

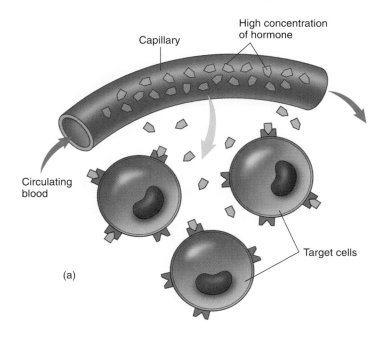

(a)

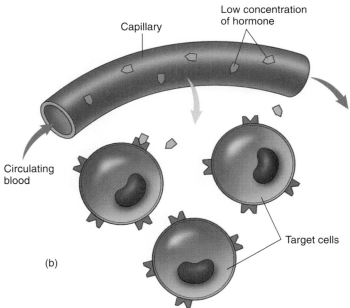

(b)

Hormone Concentration at the Target Cell
Figure 17.9

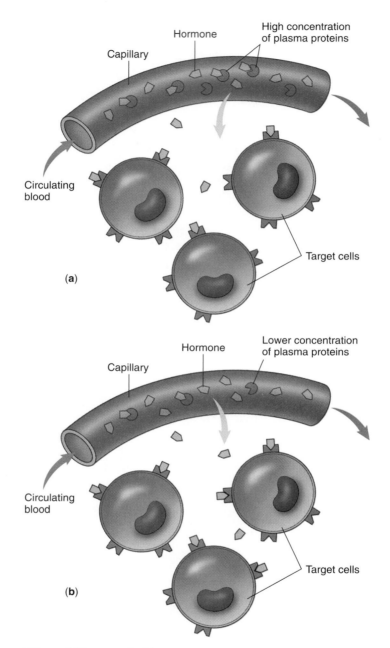

Effect of Changes in Plasma Protein Concentration on the Concentration of Free Hormone
Figure 17.10

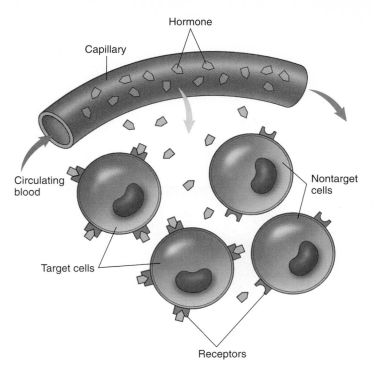

Response of Target Cells to Hormones
Figure 17.11

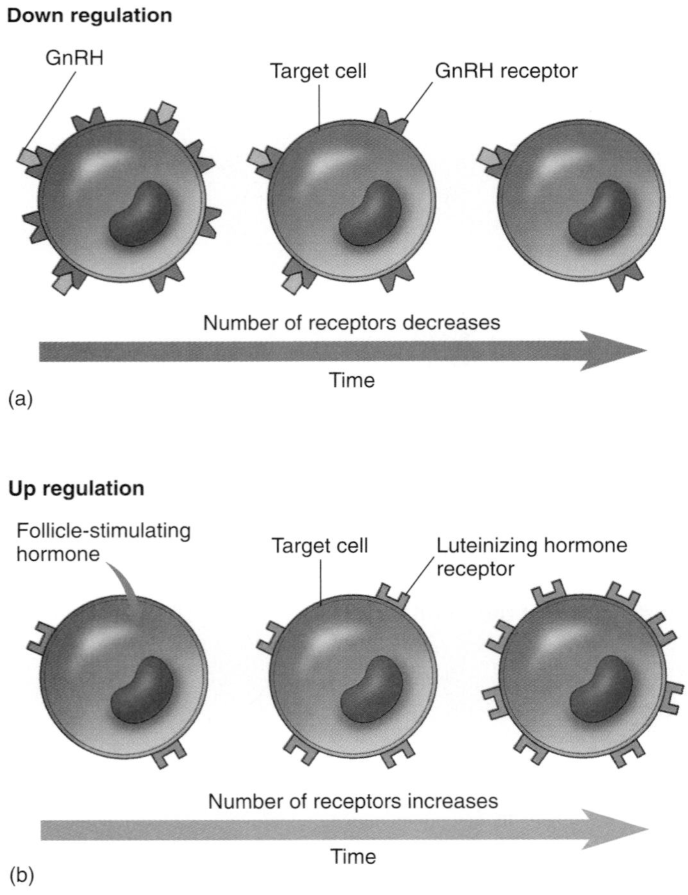

Down regulation

GnRH

Target cell   GnRH receptor

Number of receptors decreases

Time

(a)

Up regulation

Follicle-stimulating
hormone

Target cell   Luteinizing hormone
receptor

Number of receptors increases

Time

(b)

Down Regulation and Up Regulation
Figure 17.12

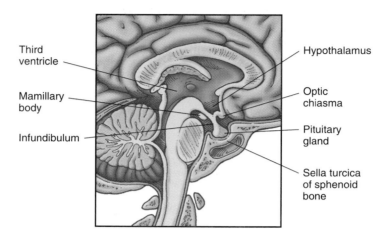

The Hypothalamus and Pituitary Gland
Figure 18.1

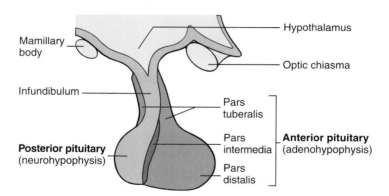

Subdivisions of the Pituitary Gland
Figure 18.2

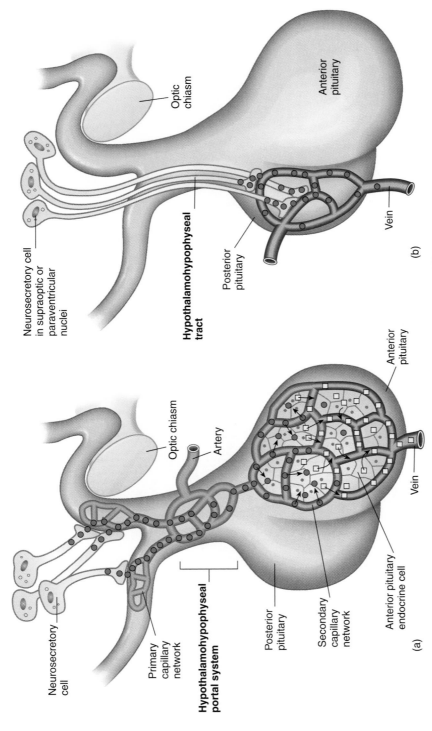

Neurosecretory cell
in supraoptic or
paraventricular
nuclei

Optic
chiasm

Anterior
pituitary

**Hypothalamohypophyseal
tract**

Posterior
pituitary

Vein

(b)

Neurosecretory
cell

Optic chiasm

Artery

Primary
capillary
network

**Hypothalamohypophyseal
portal system**

Posterior
pituitary

Secondary
capillary
network

Anterior
pituitary

Anterior pituitary
endocrine cell

Vein

(a)

The Hypothalamohypophyseal Portal System and the Hypothalamohypophyseal Tract
Figure 18.3

193

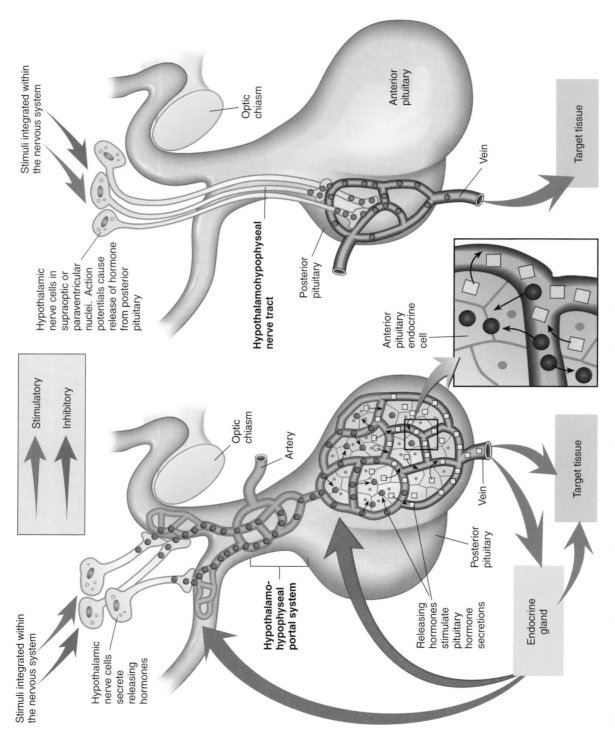

General Relationship Between the Hypothalamus, the Pituitary, and Target Tissues
Figure 18.4

Top diagram labels:

Stimuli integrated within the nervous system

Optic chiasm

Anterior pituitary

Vein

Target tissue

Hypothalamic nerve cells in supraoptic or paraventricular nuclei. Action potentials cause release of hormone from posterior pituitary

Posterior pituitary

Hypothalamohypophyseal nerve tract

Bottom diagram labels:

Stimuli integrated within the nervous system

Stimulatory

Inhibitory

Optic chiasm

Artery

Hypothalamic nerve cells secrete releasing hormones

Hypothalamo-hypophyseal portal system

Anterior pituitary endocrine cell

Releasing hormones stimulate pituitary hormone secretions

Posterior pituitary

Vein

Target tissue

Endocrine gland

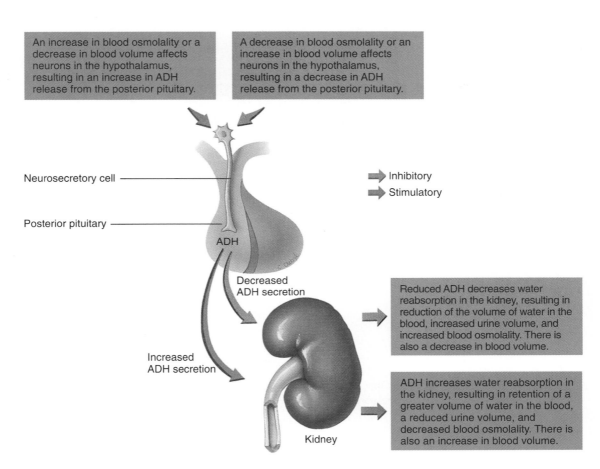

An increase in blood osmolality or a decrease in blood volume affects neurons in the hypothalamus, resulting in an increase in ADH release from the posterior pituitary.

A decrease in blood osmolality or an increase in blood volume affects neurons in the hypothalamus, resulting in a decrease in ADH release from the posterior pituitary.

Neurosecretory cell

Posterior pituitary

ADH

Decreased ADH secretion

Increased ADH secretion

Inhibitory

Stimulatory

Reduced ADH decreases water reabsorption in the kidney, resulting in reduction of the volume of water in the blood, increased urine volume, and increased blood osmolality. There is also a decrease in blood volume.

ADH increases water reabsorption in the kidney, resulting in retention of a greater volume of water in the blood, a reduced urine volume, and decreased blood osmolality. There is also an increase in blood volume.

Kidney

Control of Antidiuretic Hormone (ADH) Secretion
Figure 18.5

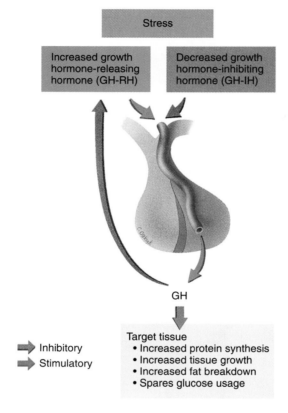

Control of Growth Hormone (GH) Secretion
Figure 18.6

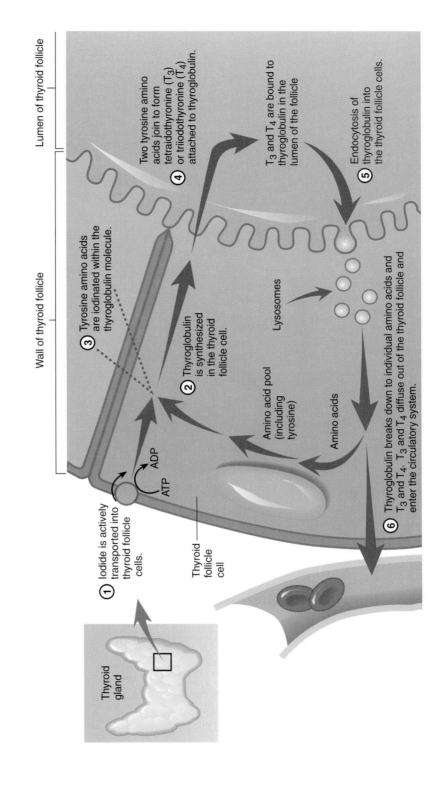

Biosynthesis of Thyroid Hormones
Figure 18.8

① Iodide is actively transported into thyroid follicle cells.

Thyroid follicle cell

ATP
ADP

② Thyroglobulin is synthesized in the thyroid follicle cell.

③ Tyrosine amino acids are iodinated within the thyroglobulin molecule.

④ Two tyrosine amino acids join to form tetraidothyronine (T_3) or triiodothyronine (T_4) attached to thyroglobulin.

⑤ Endocytosis of thyroglobulin into the thyroid follicle cells.

T_3 and T_4 are bound to thyroglobulin in the lumen of the follicle

Lysosomes

Amino acid pool (including tyrosine)

Amino acids

⑥ Thyroglobulin breaks down to individual amino acids and T_3 and T_4. T_3 and T_4 diffuse out of the thyroid follicle and enter the circulatory system.

Wall of thyroid follicle

Lumen of thyroid follicle

Thyroid gland

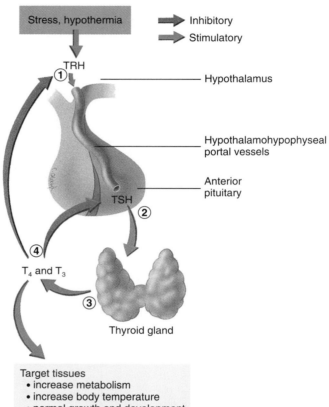

1. Thyroid-releasing hormone (TRH) is released from neurons within the hypothalamus and passes through the hypothalamohypophyseal portal blood vessels to the anterior pituitary.

2. TRH causes cells of the anterior pituitary to secrete thyroid-stimulating hormone (TSH).

3. TSH passes through the general circulation to the thyroid gland, where it causes both increased synthesis and secretion of thyroid hormones (T_3 and T_4).

4. T_3 and T_4 have an inhibitory effect on the secretion of TRH from the hypothalamus and TSH from the anterior pituitary.

Stress, hypothermia

➡ Inhibitory

➡ Stimulatory

TRH

① Hypothalamus

Hypothalamohypophyseal portal vessels

Anterior pituitary

TSH ②

④

T_4 and T_3

③

Thyroid gland

Target tissues
- increase metabolism
- increase body temperature
- normal growth and development

Regulation of Thyroid Hormone (T3 and T4) Secretion
Figure 18.9

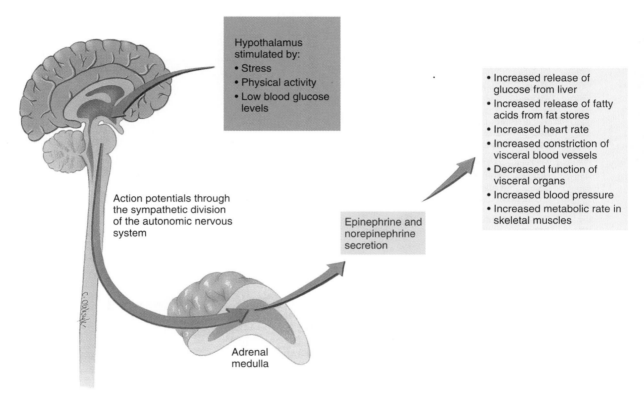

Regulation of Adrenal Medullary Secretions
Figure 18.13

Hypothalamus stimulated by:
• Stress
• Physical activity
• Low blood glucose levels

Action potentials through the sympathetic division of the autonomic nervous system

Adrenal medulla

Epinephrine and norepinephrine secretion

• Increased release of glucose from liver
• Increased release of fatty acids from fat stores
• Increased heart rate
• Increased constriction of visceral blood vessels
• Decreased function of visceral organs
• Increased blood pressure
• Increased metabolic rate in skeletal muscles

1. Cortiocotropin-releasing hormone (CRH) is released from hypothalamic neurons in response to stress or hypoglycemia and passes, by way of the hypothalamohypophyseal portal blood vessels, to the anterior pituitary.

2. In the anterior pituitary CRH binds to and stimulates cells that secrete adrenocorticotropic hormone (ACTH).

3. ACTH binds to membrane-bound receptors on cells of the adrenal cortex and stimulates the secretion of glucocorticoids, primarily cortisol.

4. Cortisol inhibits CRH and ACTH secretion.

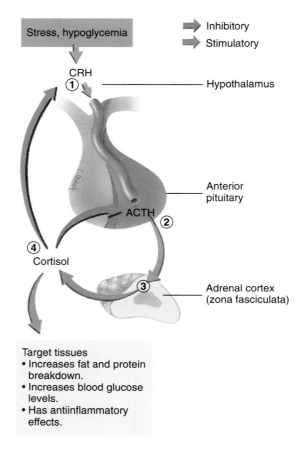

Inhibitory

Stimulatory

Stress, hypoglycemia

CRH
①

Hypothalamus

Anterior pituitary

ACTH
②

④
Cortisol

③

Adrenal cortex
(zona fasciculata)

Target tissues
• Increases fat and protein breakdown.
• Increases blood glucose levels.
• Has antiinflammatory effects.

Regulation of Cortisol Secretion
Figure 18.14

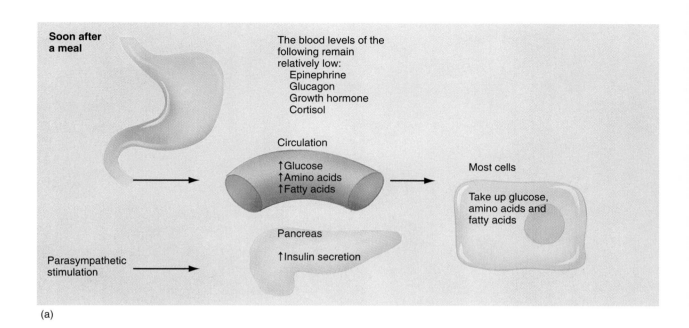

(a)

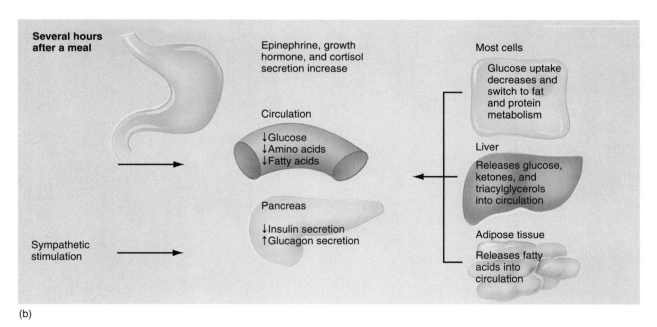

(b)

Regulation of Blood Nutrient Levels after a Meal
Figure 18.17

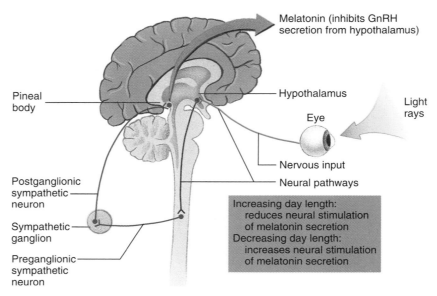

Melatonin (inhibits GnRH secretion from hypothalamus)

Pineal body

Hypothalamus

Light rays

Eye

Postganglionic sympathetic neuron

Nervous input

Neural pathways

Sympathetic ganglion

Preganglionic sympathetic neuron

Increasing day length:
 reduces neural stimulation
 of melatonin secretion
Decreasing day length:
 increases neural stimulation
 of melatonin secretion

Regulation of Pineal Secretion
Figure 18.19

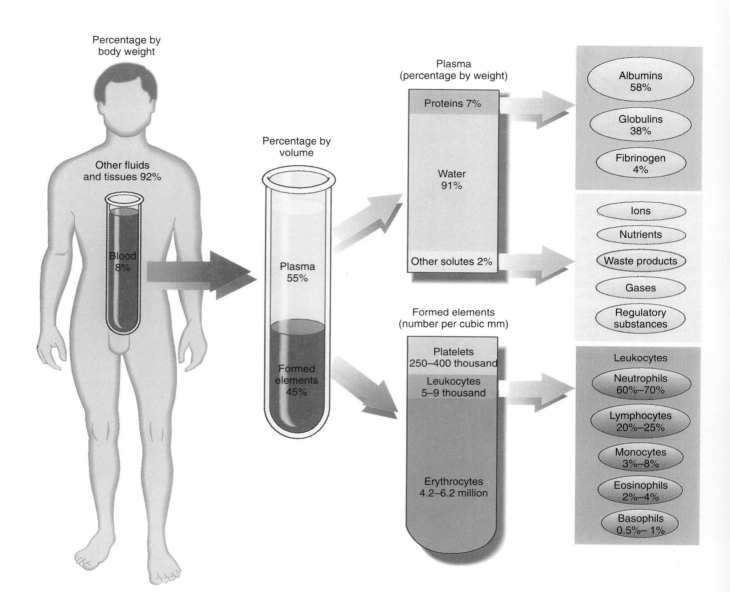

Percentage by body weight

Other fluids and tissues 92%

Blood 8%

Percentage by volume

Plasma 55%

Formed elements 45%

Plasma (percentage by weight)

Proteins 7%

Water 91%

Other solutes 2%

Albumins 58%

Globulins 38%

Fibrinogen 4%

Ions

Nutrients

Waste products

Gases

Regulatory substances

Formed elements (number per cubic mm)

Platelets 250–400 thousand

Leukocytes 5–9 thousand

Erythrocytes 4.2–6.2 million

Leukocytes

Neutrophils 60%–70%

Lymphocytes 20%–25%

Monocytes 3%–8%

Eosinophils 2%–4%

Basophils 0.5%–1%

Composition of Blood
Figure 19.1

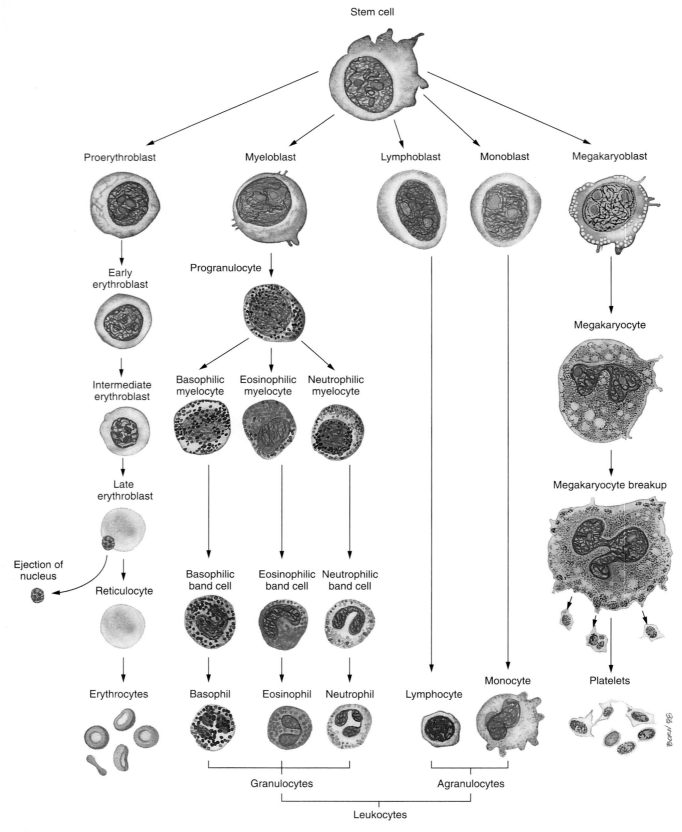

Stem cell

Proerythroblast

Early
erythroblast

Intermediate
erythroblast

Late
erythroblast

Ejection of
nucleus

Reticulocyte

Erythrocytes

Myeloblast

Progranulocyte

Basophilic
myelocyte

Eosinophilic
myelocyte

Neutrophilic
myelocyte

Basophilic
band cell

Eosinophilic
band cell

Neutrophilic
band cell

Basophil

Eosinophil

Neutrophil

Granulocytes

Lymphoblast

Monoblast

Lymphocyte

Monocyte

Agranulocytes

Leukocytes

Megakaryoblast

Megakaryocyte

Megakaryocyte breakup

Platelets

Hematopoiesis
Figure 19.2

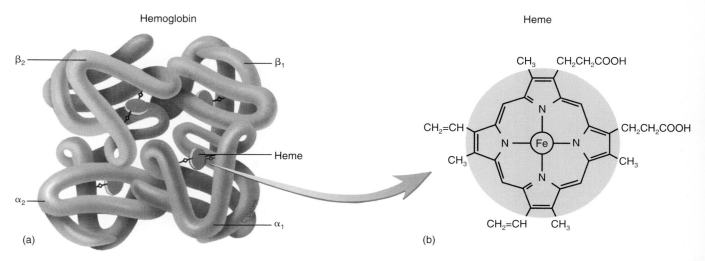

Hemoglobin

Hemoglobin
Figure 19.4

(a)

(b)

Heme

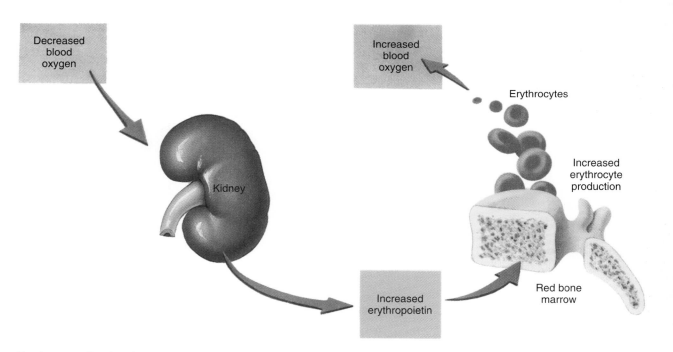

Erythrocyte Production
Figure 19.5

1. Globin chains are broken down to individual amino acids (*lavender arrow*) and are metabolized or used to build new proteins.

2. Heme is broken down to iron and bilirubin.

3. Iron is distributed to various tissues for storage or transported to the red bone marrow and used in the production of new hemoglobin (*green arrows*).

4. Free bilirubin is transported to the liver.

5. Most conjugated bilirubin is excreted as part of the bile; some is excreted in the urine.

6. Bilirubin derivatives contribute to the color of feces or are reabsorbed and excreted in the urine.

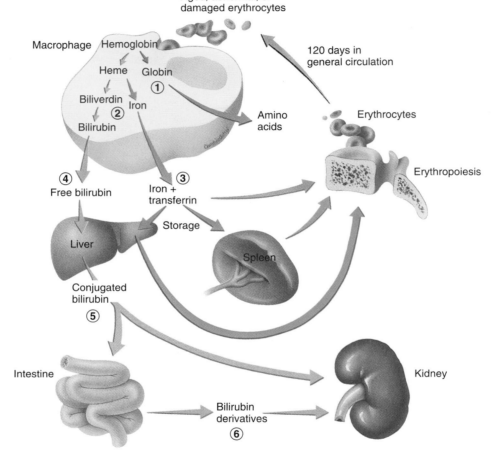

Hemoglobin Breakdown
Figure 19.6

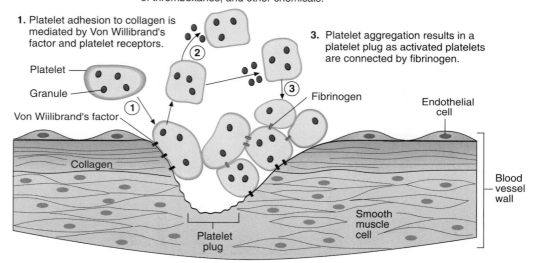

2. The platelet release reaction activates additional platelets through the release of thromboxanes, and other chemicals.

1. Platelet adhesion to collagen is mediated by Von Willibrand's factor and platelet receptors.

3. Platelet aggregation results in a platelet plug as activated platelets are connected by fibrinogen.

Platelet

Granule

Von Wiilibrand's factor

Collagen

Fibrinogen

Endothelial cell

Blood vessel wall

Platelet plug

Smooth muscle cell

Platelet Plug Formation
Figure 19.9

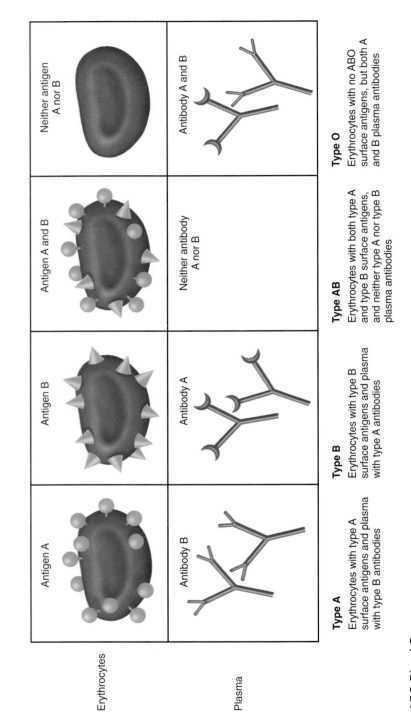

ABO Blood Groups
Figure 19.12

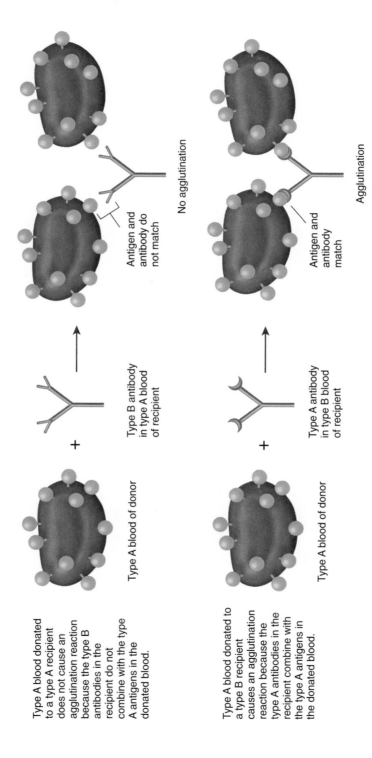

Type A blood donated to a type A recipient does not cause an agglutination reaction because the type B antibodies in the recipient do not combine with the type A antigens in the donated blood.

Type A blood of donor

+

Type B antibody in type A blood of recipient

Antigen and antibody do not match

No agglutination

Type A blood donated to a type B recipient causes an agglutination reaction because the type A antibodies in the recipient combine with the type A antigens in the donated blood.

Type A blood of donor

+

Type A antibody in type B blood of recipient

Antigen and antibody match

Agglutination

Agglutination Reaction
Figure 19.13

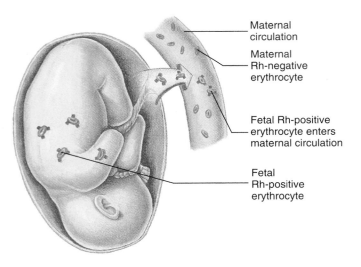

Maternal circulation

Maternal Rh-negative erythrocyte

Fetal Rh-positive erythrocyte enters maternal circulation

Fetal Rh-positive erythrocyte

1. Before or during delivery, Rh-positive erythrocytes from the fetus enter the blood of an Rh-negative woman through a tear in the placenta.

(a)

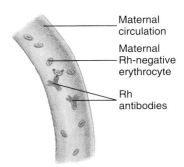

Maternal circulation

Maternal Rh-negative erythrocyte

Rh antibodies

2. The mother is sensitized to the Rh antigen and produces Rh antibodies. Because this usually happens after delivery, there is no effect on the fetus in the first pregnancy.

(b)

3. During a subsequent pregnancy with an Rh-positive fetus, Rh-positive erythrocytes cross the placenta, enter the maternal circulation, and stimulate the mother to produce antibodies against the Rh antigen. Antibody production is rapid because the mother has been sensitized to the Rh antigen. The Rh antibodies from the mother cross the placenta, causing agglutination and hemolysis of fetal erythrocytes, and hemolytic disease of the newborn develops.

Maternal circulation

Maternal Rh antibodies cross the placenta

Agglutination of fetal Rh-positive erythrocytes leads to hemolytic disease of the newborn

(c)

Hemolytic Disease of the Newborn (HDN)
Figure 19.14

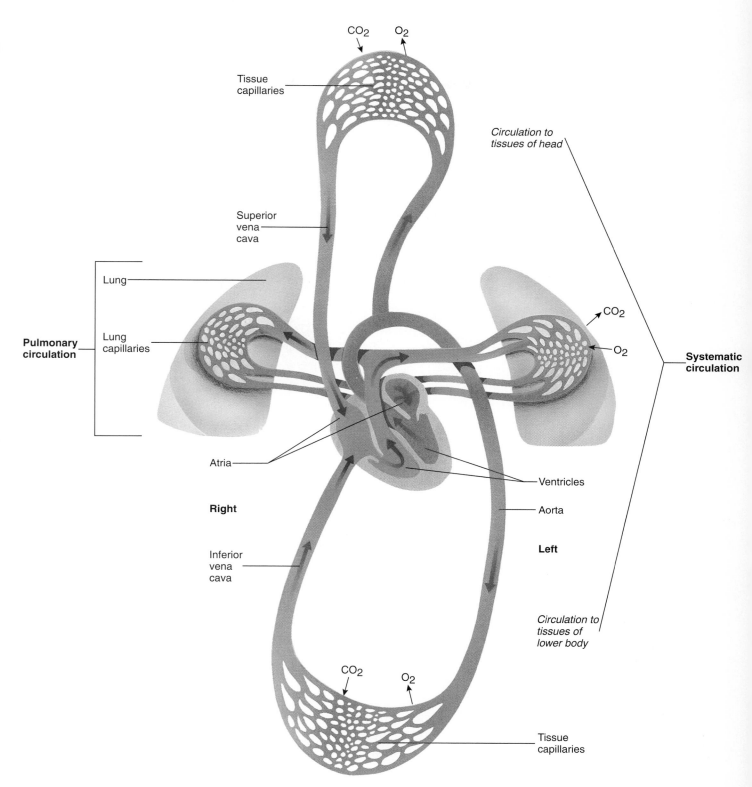

CO_2 O_2

Tissue capillaries

Circulation to tissues of head

Superior vena cava

Lung

Lung capillaries

Pulmonary circulation

CO_2

O_2

Systematic circulation

Atria

Ventricles

Aorta

Right

Left

Inferior vena cava

Circulation to tissues of lower body

CO_2 O_2

Tissue capillaries

The Systemic and Pulmonary Circulation
Figure 20.1

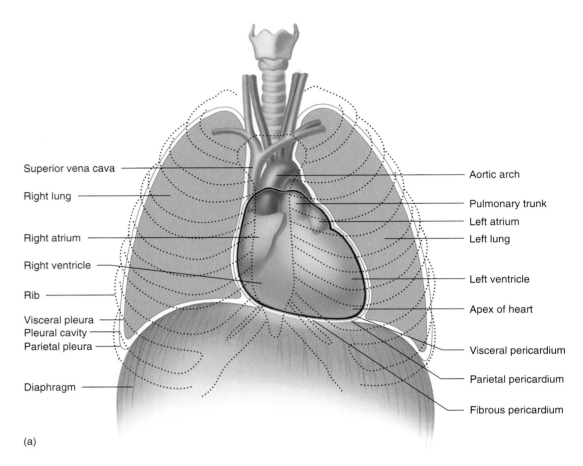

Superior vena cava

Right lung

Right atrium

Right ventricle

Rib

Visceral pleura
Pleural cavity
Parietal pleura

Diaphragm

Aortic arch

Pulmonary trunk
Left atrium
Left lung

Left ventricle

Apex of heart

Visceral pericardium

Parietal pericardium

Fibrous pericardium

(a)

Location of the Heart in the Thorax
Figure 20.2a

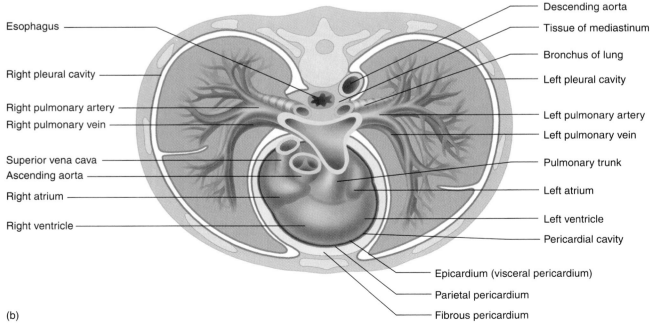

Esophagus

Right pleural cavity

Right pulmonary artery
Right pulmonary vein

Superior vena cava
Ascending aorta
Right atrium

Right ventricle

Descending aorta

Tissue of mediastinum

Bronchus of lung

Left pleural cavity

Left pulmonary artery
Left pulmonary vein

Pulmonary trunk

Left atrium

Left ventricle
Pericardial cavity

Epicardium (visceral pericardium)

Parietal pericardium

Fibrous pericardium

(b)

Cross Section of the Thorax
Figure 20.2b

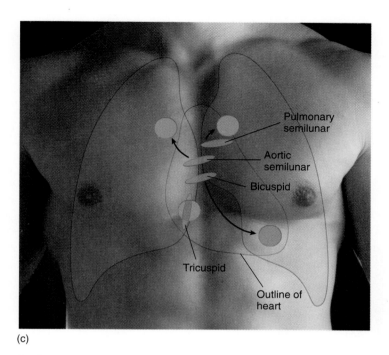

(c)

Surface Markings of the Heart in the Male
Figure 20.2c

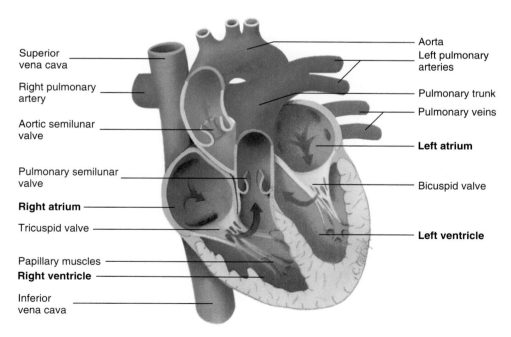

Chambers of the Heart
Figure 20.3

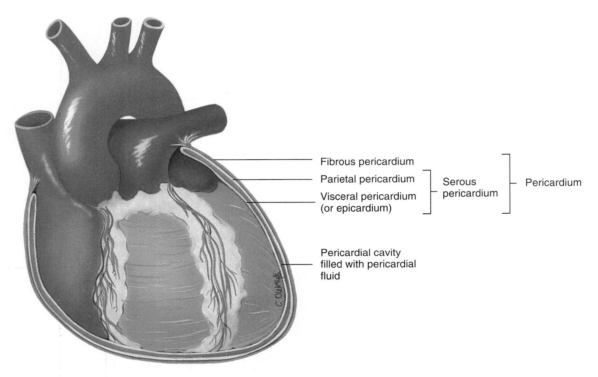

Fibrous pericardium

Parietal pericardium

Visceral pericardium
(or epicardium)

Serous
pericardium

Pericardium

Pericardial cavity
filled with pericardial
fluid

The Pericardium
Figure 20.4

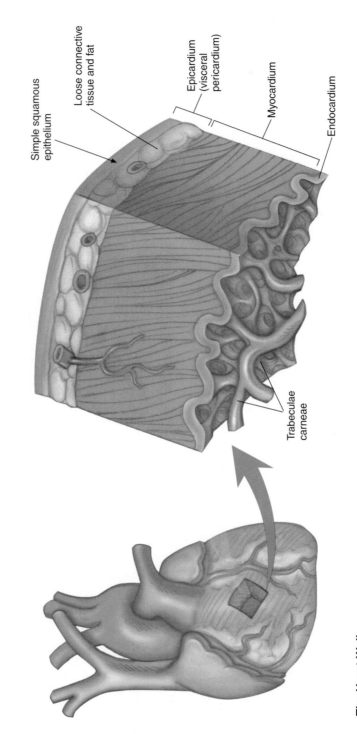

Simple squamous epithelium

Loose connective tissue and fat

Epicardium (visceral pericardium)

Myocardium

Endocardium

Trabeculae carneae

The Heart Wall
Figure 20.5

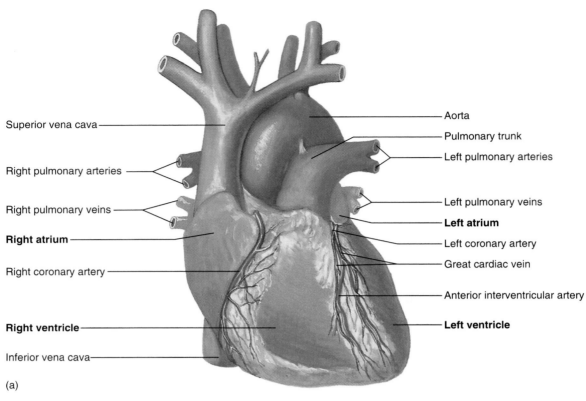

Superior vena cava

Right pulmonary arteries

Right pulmonary veins

Right atrium

Right coronary artery

Right ventricle

Inferior vena cava

Aorta

Pulmonary trunk

Left pulmonary arteries

Left pulmonary veins

Left atrium

Left coronary artery

Great cardiac vein

Anterior interventricular artery

Left ventricle

(a)

Surface of the Heart (Anterior)
Figure 20.6a

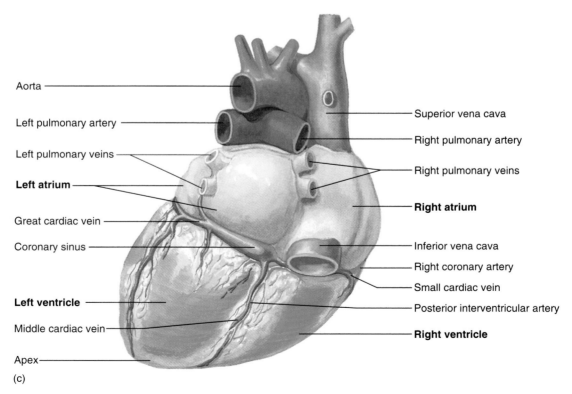

Aorta

Left pulmonary artery

Left pulmonary veins

Left atrium

Great cardiac vein

Coronary sinus

Left ventricle

Middle cardiac vein

Apex

Superior vena cava

Right pulmonary artery

Right pulmonary veins

Right atrium

Inferior vena cava

Right coronary artery

Small cardiac vein

Posterior interventricular artery

Right ventricle

(c)

Surface of the Heart (Posterior)
Figure 20.6c

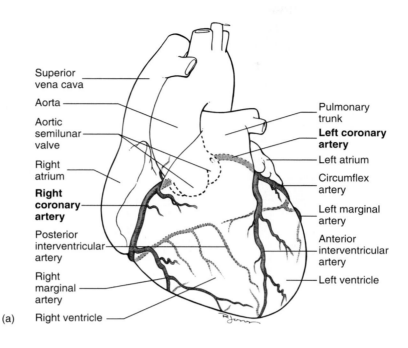

Superior vena cava

Aorta

Aortic semilunar valve

Right atrium

Right coronary artery

Posterior interventricular artery

Right marginal artery

(a)   Right ventricle

Pulmonary trunk

Left coronary artery

Left atrium

Circumflex artery

Left marginal artery

Anterior interventricular artery

Left ventricle

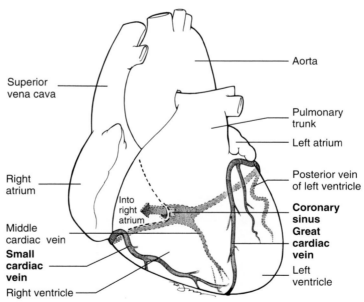

Superior vena cava

Right atrium

Middle cardiac vein

Small cardiac vein

Right ventricle

(b)

Aorta

Pulmonary trunk

Left atrium

Posterior vein of left ventricle

Coronary sinus Great cardiac vein

Left ventricle

Into right atrium

Arteries Supplying the Heart and Veins Draining the Heart
Figure 20.7

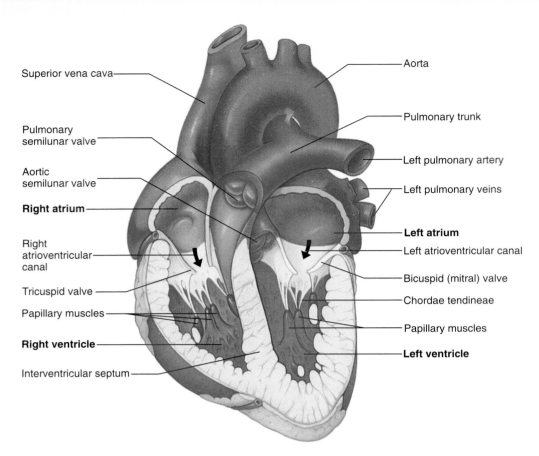

Superior vena cava

Pulmonary
semilunar valve

Aortic
semilunar valve

Right atrium

Right
atrioventricular
canal

Tricuspid valve

Papillary muscles

Right ventricle

Interventricular septum

Aorta

Pulmonary trunk

Left pulmonary artery

Left pulmonary veins

Left atrium

Left atrioventricular canal

Bicuspid (mitral) valve

Chordae tendineae

Papillary muscles

Left ventricle

Internal Anatomy of the Heart
Figure 20.8

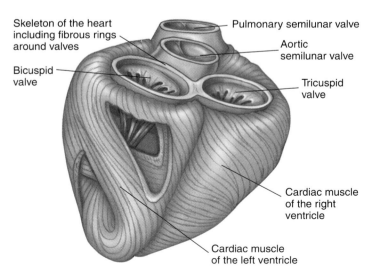

Skeleton of the heart
including fibrous rings
around valves

Bicuspid
valve

Pulmonary semilunar valve

Aortic
semilunar valve

Tricuspid
valve

Cardiac muscle
of the right
ventricle

Cardiac muscle
of the left ventricle

Skeleton of the Heart
Figure 20.10

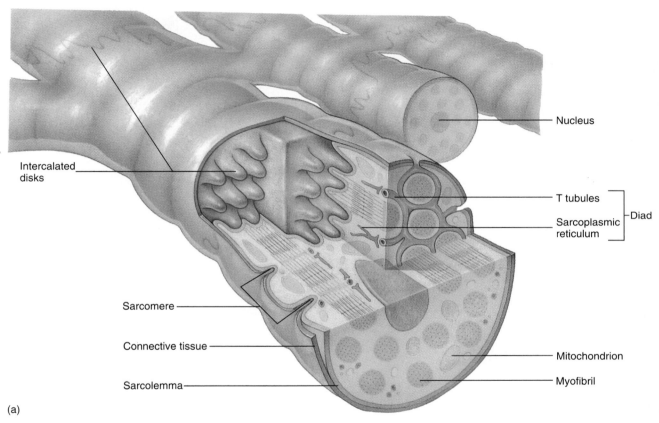

Intercalated disks

Nucleus

T tubules

Sarcoplasmic reticulum

Diad

Sarcomere

Connective tissue

Sarcolemma

Mitochondrion

Myofibril

(a)

Histology of the Heart
Figure 20.11a

1. Action potentials originate in the sinoatrial (SA) node and travel across the wall of the atrium (arrows) from the SA node to the atrioventricular (AV) node.

2. Action potentials pass through the AV node and along the atrioventricular (AV) bundle, which extends from the AV node, through the fibrous skeleton, into the interventricular septum.

3. The AV bundle divides into right and left bundle branches, and action potentials descend to the apex of each ventricle along the bundle branches.

4. Action potentials are carried by the Purkinje fibers from the bundle branches to the ventricular walls.

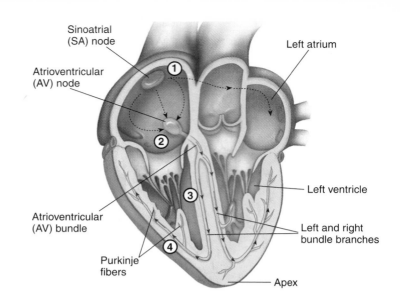

Conducting System of the Heart
Figure 20.12

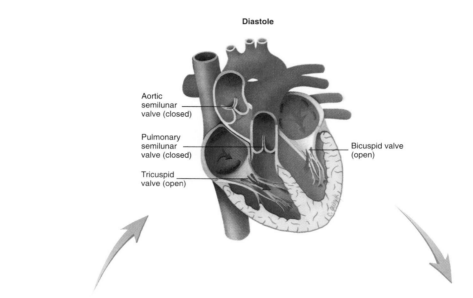

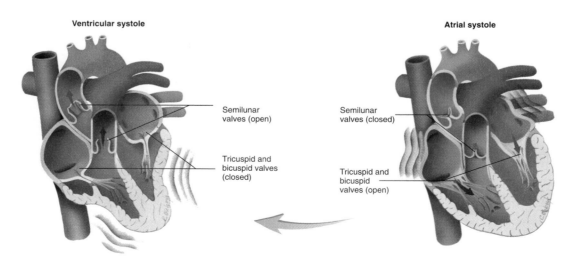

The Cardiac Cycle
Figure 20.16

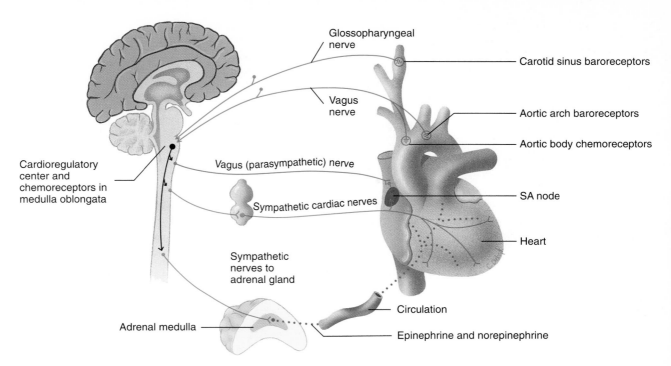

Regulation of the Heart
Figure 20.20

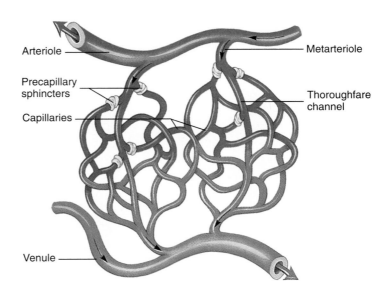

Capillary Network
Figure 21.2

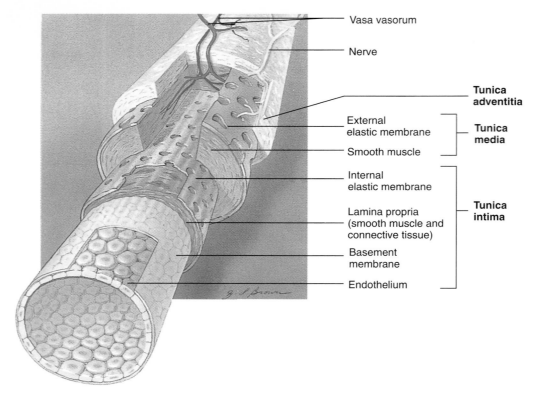

Histology of a Blood Vessel
Figure 21.3

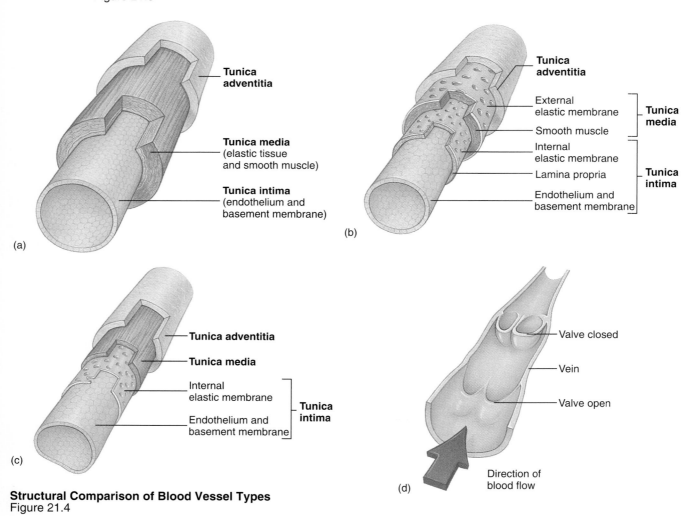

(a)

(b)

(c)

(d)

Structural Comparison of Blood Vessel Types
Figure 21.4

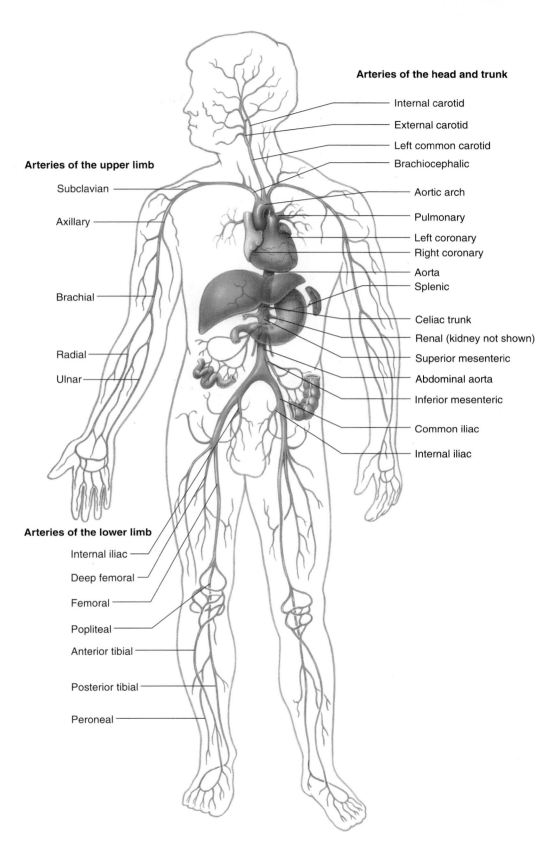

Arteries of the head and trunk

Internal carotid
External carotid
Left common carotid
Brachiocephalic
Aortic arch
Pulmonary
Left coronary
Right coronary
Aorta
Splenic
Celiac trunk
Renal (kidney not shown)
Superior mesenteric
Abdominal aorta
Inferior mesenteric
Common iliac
Internal iliac

Arteries of the upper limb

Subclavian
Axillary
Brachial
Radial
Ulnar

Arteries of the lower limb

Internal iliac
Deep femoral
Femoral
Popliteal
Anterior tibial
Posterior tibial
Peroneal

The Major Arteries
Figure 21.6

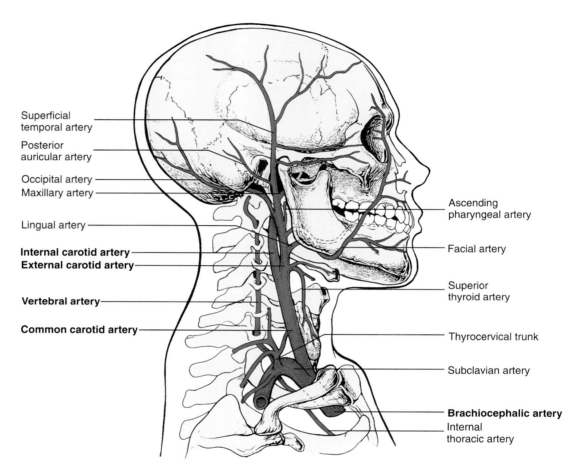

Superficial
temporal artery

Posterior
auricular artery

Occipital artery

Maxillary artery

Lingual artery

Internal carotid artery

External carotid artery

Vertebral artery

Common carotid artery

Ascending
pharyngeal artery

Facial artery

Superior
thyroid artery

Thyrocervical trunk

Subclavian artery

Brachiocephalic artery

Internal
thoracic artery

Arteries of the Head and Neck
Figure 21.7

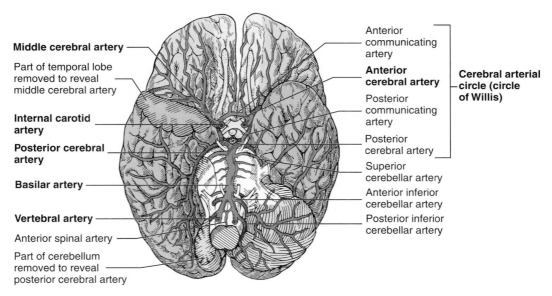

Middle cerebral artery

Part of temporal lobe
removed to reveal
middle cerebral artery

Internal carotid
artery

Posterior cerebral
artery

Basilar artery

Vertebral artery

Anterior spinal artery

Part of cerebellum
removed to reveal
posterior cerebral artery

Anterior
communicating
artery

Anterior
cerebral artery

Posterior
communicating
artery

Posterior
cerebral artery

Superior
cerebellar artery

Anterior inferior
cerebellar artery

Posterior inferior
cerebellar artery

Cerebral arterial
circle (circle
of Willis)

Inferior View of the Brain Showing Arteries
Figure 21.8

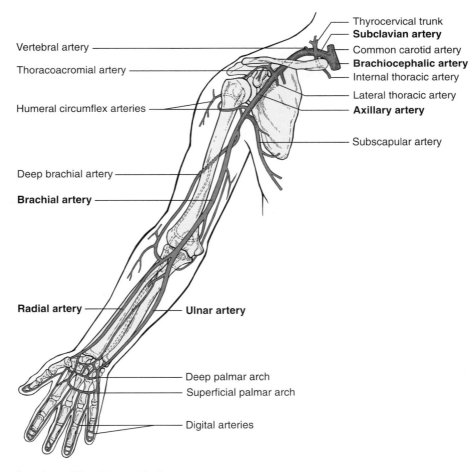

Vertebral artery

Thoracoacromial artery

Humeral circumflex arteries

Deep brachial artery

Brachial artery

Radial artery

Ulnar artery

Thyrocervical trunk

Subclavian artery

Common carotid artery

Brachiocephalic artery

Internal thoracic artery

Lateral thoracic artery

Axillary artery

Subscapular artery

Deep palmar arch

Superficial palmar arch

Digital arteries

Arteries of the Upper Limb
Figure 21.10

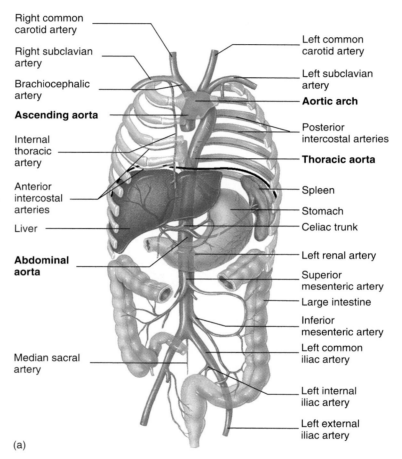

Right common
carotid artery

Right subclavian
artery

Brachiocephalic
artery

Ascending aorta

Internal
thoracic
artery

Anterior
intercostal
arteries

Liver

**Abdominal
aorta**

Median sacral
artery

Left common
carotid artery

Left subclavian
artery

Aortic arch

Posterior
intercostal arteries

Thoracic aorta

Spleen

Stomach

Celiac trunk

Left renal artery

Superior
mesenteric artery

Large intestine

Inferior
mesenteric artery

Left common
iliac artery

Left internal
iliac artery

Left external
iliac artery

(a)

Branches of the Aorta
Figure 21.12a

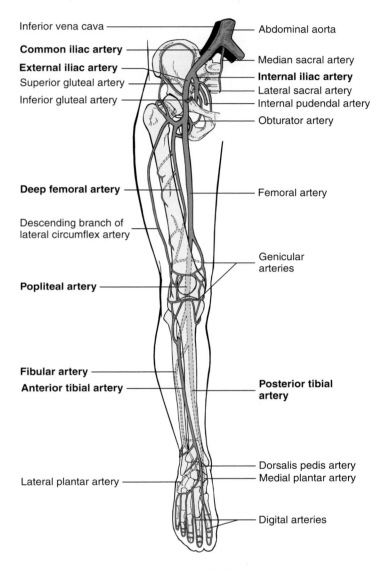

Inferior vena cava

Common iliac artery

External iliac artery

Superior gluteal artery

Inferior gluteal artery

Deep femoral artery

Descending branch of
lateral circumflex artery

Popliteal artery

Fibular artery

Anterior tibial artery

Lateral plantar artery

Abdominal aorta

Median sacral artery

Internal iliac artery

Lateral sacral artery

Internal pudendal artery

Obturator artery

Femoral artery

Genicular
arteries

**Posterior tibial
artery**

Dorsalis pedis artery

Medial plantar artery

Digital arteries

Arteries of the Pelvis and Lower Limb
Figure 21.14

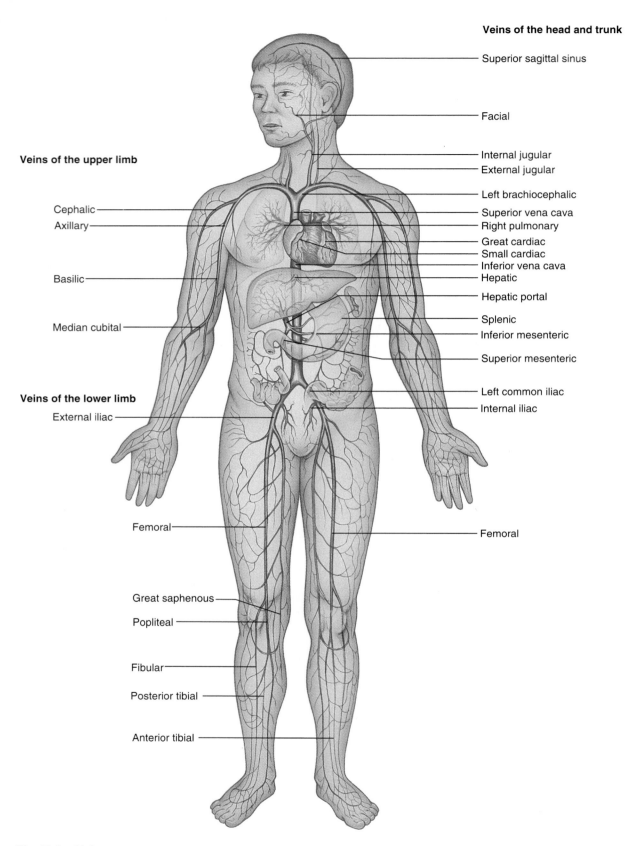

Veins of the head and trunk

Superior sagittal sinus

Facial

Internal jugular
External jugular

Left brachiocephalic
Superior vena cava
Right pulmonary
Great cardiac
Small cardiac
Inferior vena cava
Hepatic
Hepatic portal
Splenic
Inferior mesenteric
Superior mesenteric

Left common iliac
Internal iliac

Femoral

Veins of the upper limb

Cephalic
Axillary

Basilic

Median cubital

Veins of the lower limb

External iliac

Femoral

Great saphenous
Popliteal

Fibular

Posterior tibial

Anterior tibial

The Major Veins
Figure 21.16

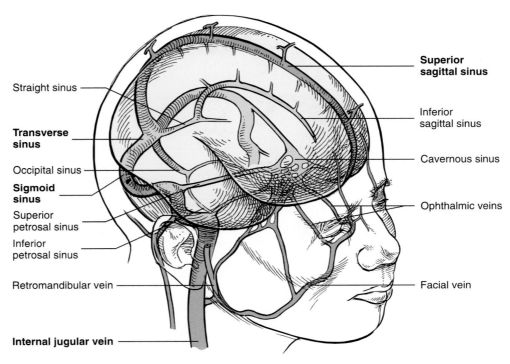

Straight sinus

Transverse sinus

Occipital sinus

Sigmoid sinus

Superior petrosal sinus

Inferior petrosal sinus

Retromandibular vein

Internal jugular vein

Superior sagittal sinus

Inferior sagittal sinus

Cavernous sinus

Ophthalmic veins

Facial vein

Venous Sinuses Associated with the Brain
Figure 21.17

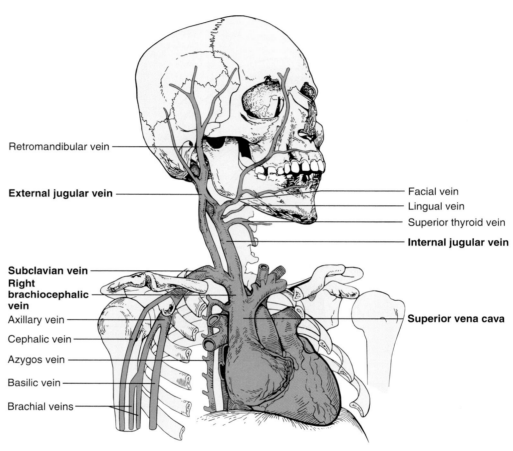

Retromandibular vein

External jugular vein

Facial vein
Lingual vein
Superior thyroid vein
Internal jugular vein

Subclavian vein
Right brachiocephalic vein

Axillary vein

Cephalic vein

Azygos vein

Basilic vein

Brachial veins

Superior vena cava

Veins of the Head and Neck
Figure 21.18

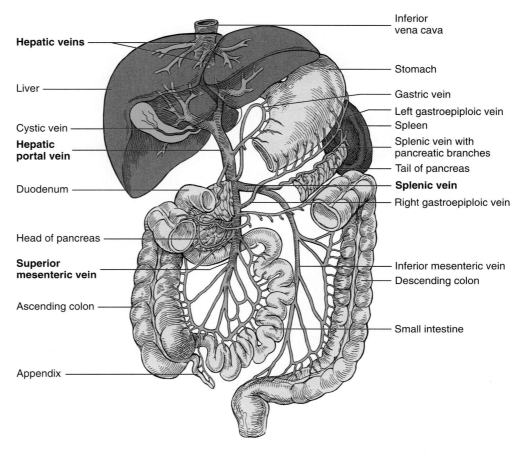

Inferior vena cava

Hepatic veins

Liver

Cystic vein

Hepatic portal vein

Duodenum

Head of pancreas

Superior mesenteric vein

Ascending colon

Appendix

Stomach

Gastric vein

Left gastroepiploic vein

Spleen

Splenic vein with pancreatic branches

Tail of pancreas

Splenic vein

Right gastroepiploic vein

Inferior mesenteric vein

Descending colon

Small intestine

Hepatic Portal System
Figure 21.24

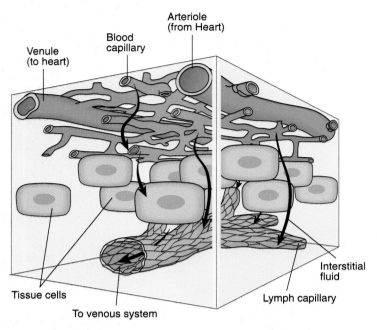

Movement of Fluid Between Blood Capillary to Lymphatic Capillaries
Figure 21.30

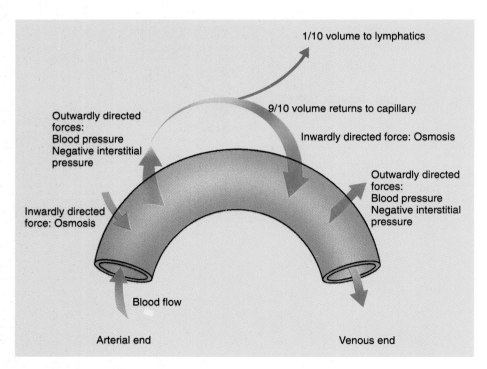

Total Pressure Differences Between the Inside and Outside of the Capillary
Figure 21.36

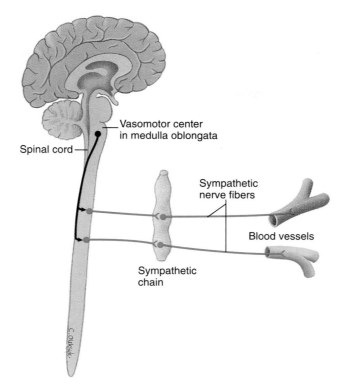

Vasomotor center
in medulla oblongata

Spinal cord

Sympathetic
nerve fibers

Blood vessels

Sympathetic
chain

Blood Vessel Innervation by Sympathetic Nerve Fibers
Figure 21.38

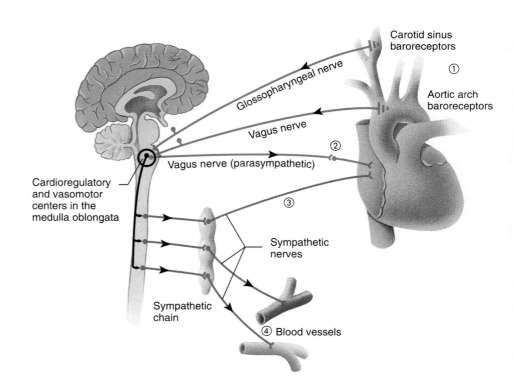

1. Baroreceptors in the carotid
sinus and aortic arch monitor
blood pressure.

2. Increased parasympathetic
stimulation of the heart
decreases the heart rate.

3. Increased sympathetic
stimulation of the heart
increases the heart rate
and stroke volume.

4. Increased sympathetic
stimulation of blood vessels
increases vasoconstriction.

Carotid sinus
baroreceptors

①

Aortic arch
baroreceptors

Glossopharyngeal nerve

Vagus nerve

Vagus nerve (parasympathetic)

②

③

Cardioregulatory
and vasomotor
centers in the
medulla oblongata

Sympathetic
nerves

Sympathetic
chain

④ Blood vessels

Baroreceptor Reflex Control of Blood Pressure
Figure 21.39

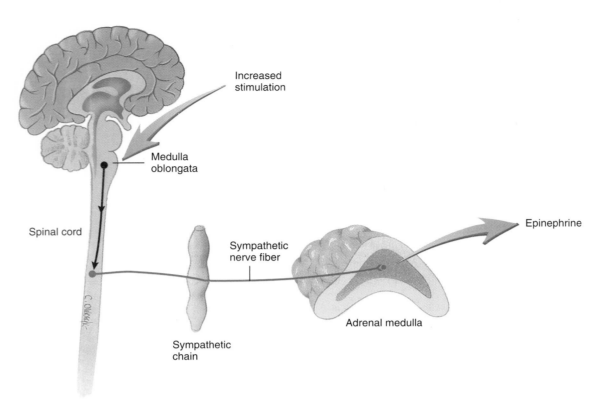

Hormonal Regulation of Blood Pressure: The Adrenal Medullary Mechanism
Figure 21.40

1. Chemoreceptors in the carotid sinus and aortic bodies monitor blood O_2, CO_2, and pH.

2. Chemoreceptors in the medulla oblongata monitor blood CO_2 and pH.

3. Increased parasympathetic stimulation of the heart decreases the heart rate.

4. Increased sympathetic stimulation of the heart increases the heart rate and stroke volume.

5. Increased sympathetic stimulation of blood vessels increases vasoconstriction.

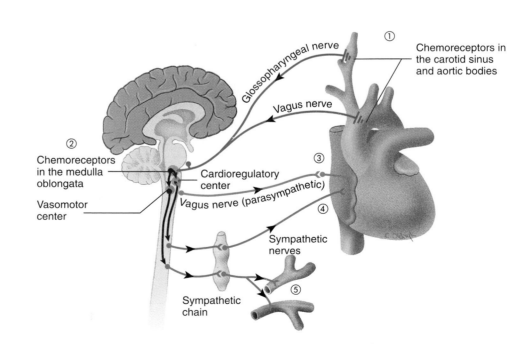

Chemoreceptor Reflex Control of Blood Pressure
Figure 21.41

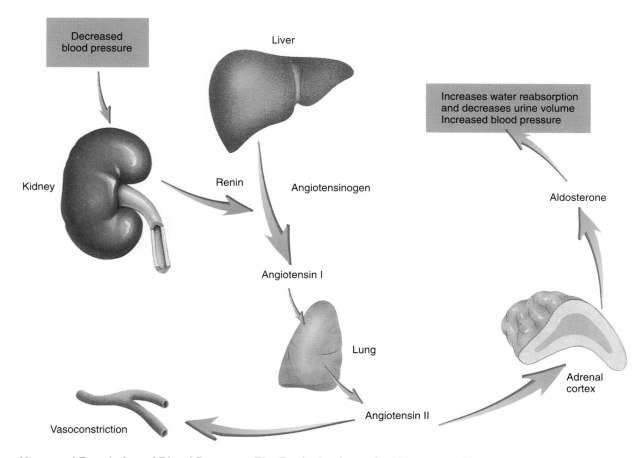

Decreased blood pressure

Liver

Increases water reabsorption and decreases urine volume Increased blood pressure

Kidney

Renin

Angiotensinogen

Aldosterone

Angiotensin I

Lung

Adrenal cortex

Vasoconstriction

Angiotensin II

Hormonal Regulation of Blood Pressure: The Renin-Angiotensin-Aldosterone Mechanism
Figure 21.44

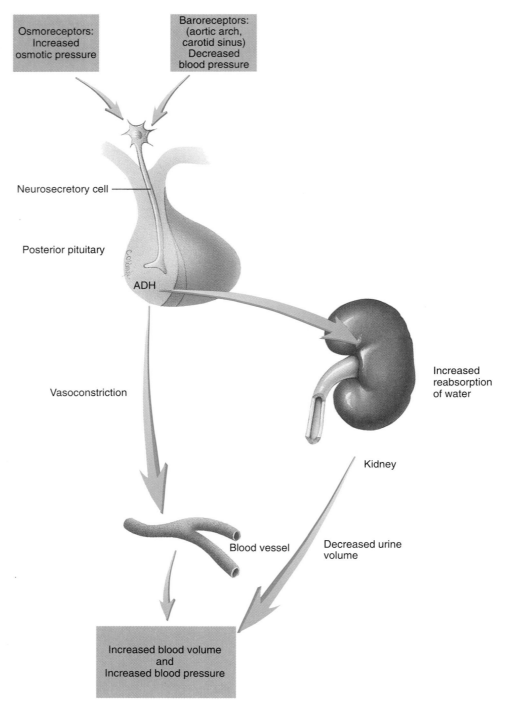

Osmoreceptors:
Increased
osmotic pressure

Baroreceptors:
(aortic arch,
carotid sinus)
Decreased
blood pressure

Neurosecretory cell

Posterior pituitary

ADH

Vasoconstriction

Increased
reabsorption
of water

Kidney

Blood vessel

Decreased urine
volume

Increased blood volume
and
Increased blood pressure

Hormonal Regulation of Blood Pressure: The Vasopressin (ADH) Mechanism
Figure 21.45

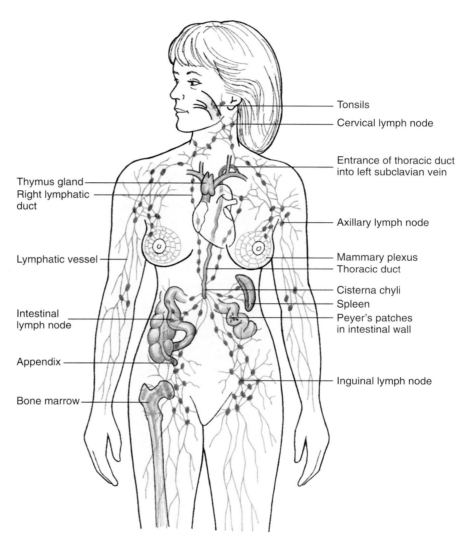

Tonsils

Cervical lymph node

Entrance of thoracic duct into left subclavian vein

Thymus gland

Right lymphatic duct

Axillary lymph node

Lymphatic vessel

Mammary plexus

Thoracic duct

Cisterna chyli

Spleen

Intestinal lymph node

Peyer's patches in intestinal wall

Appendix

Inguinal lymph node

Bone marrow

Lymphatic System
Figure 22.1

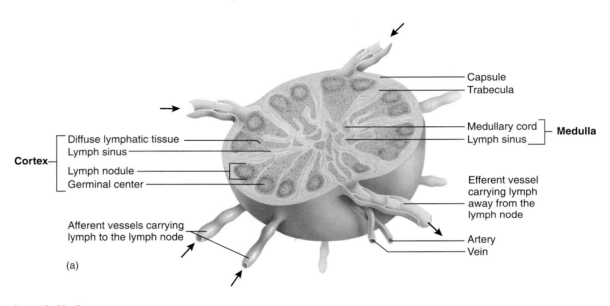

Capsule
Trabecula

Medullary cord ⎤
                ⎦ **Medulla**
Lymph sinus ⎦

Diffuse lymphatic tissue
Lymph sinus

Cortex

Lymph nodule
Germinal center

Efferent vessel
carrying lymph
away from the
lymph node

Afferent vessels carrying
lymph to the lymph node

Artery
Vein

(a)

Lymph Node
Figure 22.4a

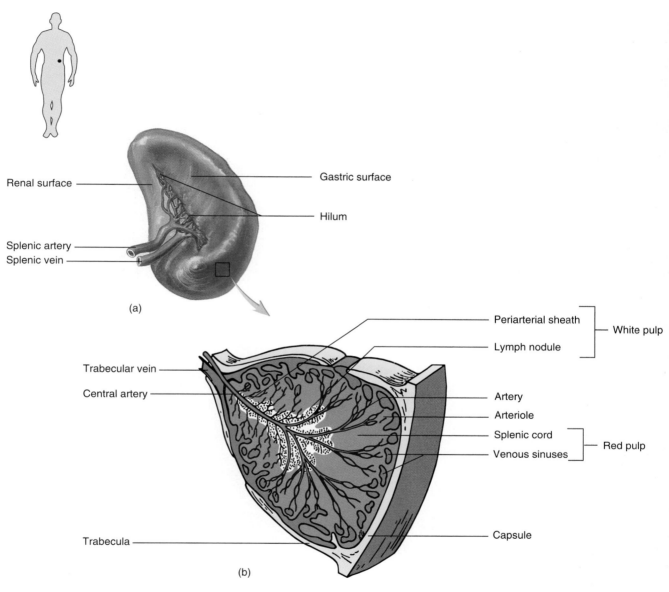

Renal surface

Gastric surface

Hilum

Splenic artery

Splenic vein

(a)

Periarterial sheath — White pulp

Lymph nodule

Trabecular vein

Central artery

Artery

Arteriole

Splenic cord

Venous sinuses — Red pulp

Capsule

Trabecula

(b)

Spleen
Figure 22.5a, b

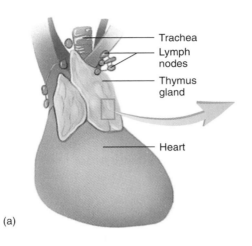

(a)

Trachea

Lymph nodes

Thymus gland

Heart

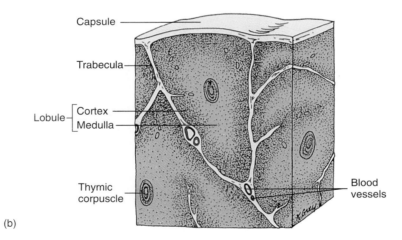

(b)

Capsule

Trabecula

Lobule — Cortex
        Medulla

Thymic corpuscle

Blood vessels

Thymus Gland
Figure 22.6a, b

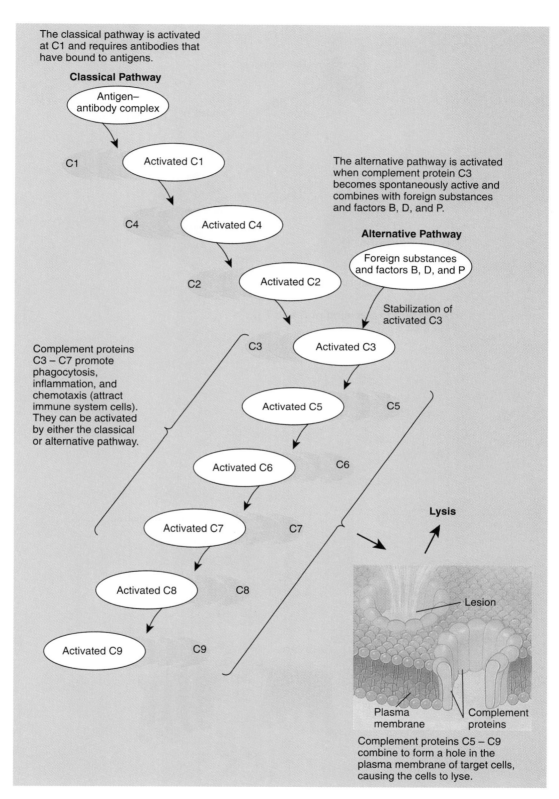

The classical pathway is activated at C1 and requires antibodies that have bound to antigens.

Classical Pathway

Antigen–antibody complex

C1

Activated C1

The alternative pathway is activated when complement protein C3 becomes spontaneously active and combines with foreign substances and factors B, D, and P.

C4

Activated C4

Alternative Pathway

Foreign substances and factors B, D, and P

C2

Activated C2

Stabilization of activated C3

Complement proteins C3 – C7 promote phagocytosis, inflammation, and chemotaxis (attract immune system cells). They can be activated by either the classical or alternative pathway.

C3

Activated C3

Activated C5

C5

Activated C6

C6

Lysis

Activated C7

C7

Activated C8

C8

Lesion

Activated C9

C9

Plasma membrane

Complement proteins

Complement proteins C5 – C9 combine to form a hole in the plasma membrane of target cells, causing the cells to lyse.

Complement Cascade
Figure 22.7

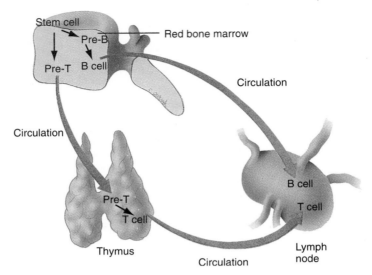

Origin and Processing of B and T Cells
Figure 22.9

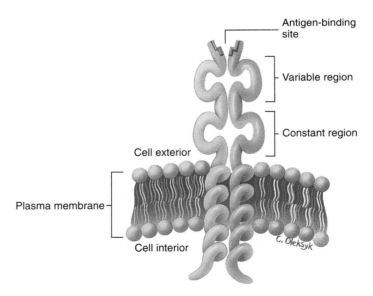

The T Cell Receptor
Figure 22.11

1. Foreign proteins or self-proteins within the cytosol are broken down into fragments that are antigens.
2. Antigens are transported into the rough endoplasmic reticulum.
3. Antigens combine with MHC class I molecules.
4. The MHC class I–antigen complex is transported to the Golgi apparatus, packaged into a vesicle, and transported to the plasma membrane.
5. Foreign antigens combined with MHC class I molecules stimulate cell destruction.
6. Self-antigens combined with MHC class I molecules do not stimulate cell destruction.

(a)

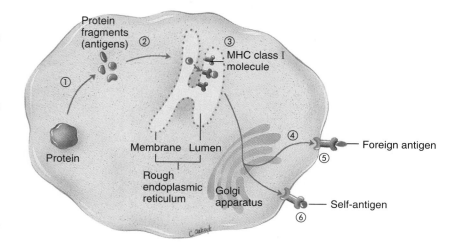

Antigen Processing
Figure 22.12a

1. The unprocessed extracellular antigen is ingested by endocytosis and is within a vesicle.
2. The antigen is broken down into fragments to form processed antigens.
3. The vesicle containing the processed antigen fuses with vesicles produced by the Golgi apparatus that contain MHC class II molecules. The processed antigen and the MHC class II molecule combine.
4. The MHC class II–antigen complex is transported to the plasma membrane.
5. The displayed MHC class II–antigen complex can stimulate immune cells.

(b)

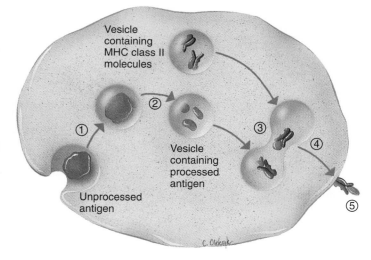

Antigen Processing
Figure 22.12b

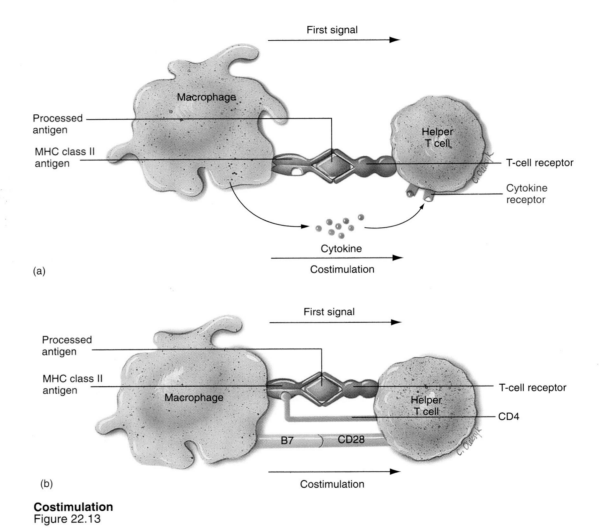

First signal

Macrophage

Processed antigen

MHC class II antigen

Helper T cell

T-cell receptor

Cytokine receptor

Cytokine

Costimulation

(a)

First signal

Processed antigen

MHC class II antigen

Macrophage

Helper T cell

T-cell receptor

CD4

B7   CD28

Costimulation

(b)

Costimulation
Figure 22.13

1. Antigen-presenting cells such as macrophages take in, process, and display antigens on the cell's surface.

2. The antigens are bound to MHC class II molecules, which function to present the processed antigen to the T cell receptor of the helper T cell for recognition.

3. Costimulation occurs by the CD4 glycoprotein of the helper T cell and by cytokines. The macrophage secretes a cytokine called interleukin-1.

4. Interleukin-1 stimulates the helper T cell to secrete the cytokine interleukin-2 and to produce interleukin-2 receptors.

5. The helper T cell stimulates itself to divide when interleukin-2 binds to interleukin-2 receptors.

6. The "daughter" helper T cells resulting from this division can be stimulated to divide again if they are exposed to the same antigen that stimulated the "parent" helper T cell. This greatly increases the number of helper T cells.

7. The increased number of helper T cells can facilitate the activation of B cells or effector T cells.

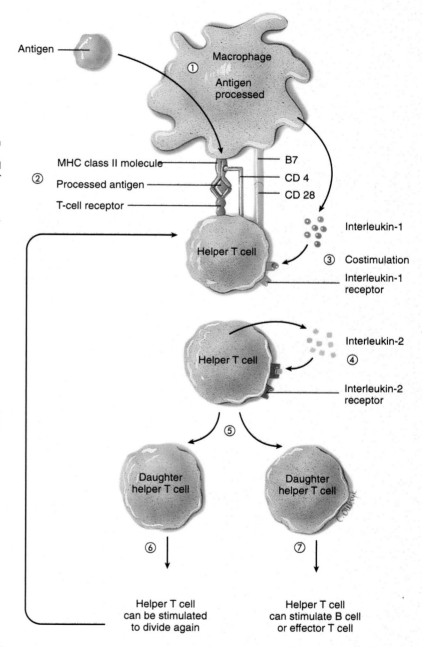

Proliferation of Helper T Cells
Figure 22.14

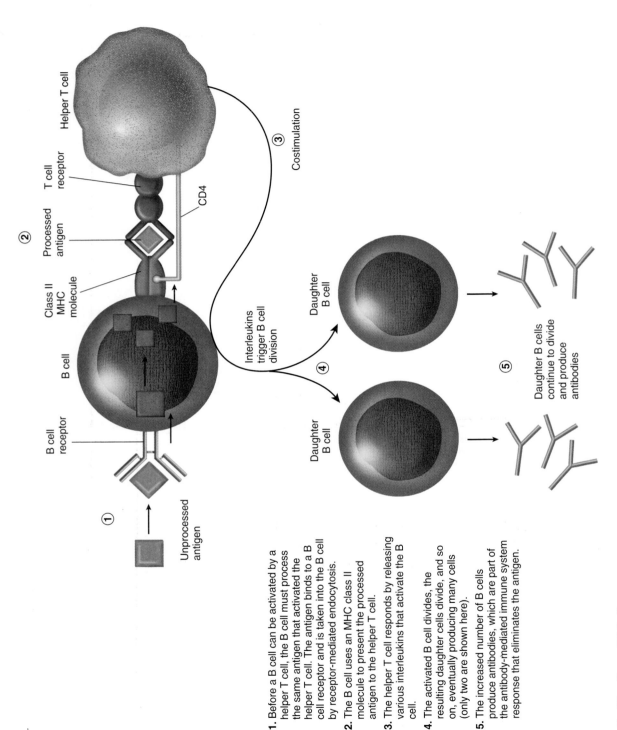

1. Before a B cell can be activated by a helper T cell, the B cell must process the same antigen that activated the helper T cell. The antigen binds to a B cell receptor and is taken into the B cell by receptor-mediated endocytosis.

2. The B cell uses an MHC class II molecule to present the processed antigen to the helper T cell.

3. The helper T cell responds by releasing various interleukins that activate the B cell.

4. The activated B cell divides, the resulting daughter cells divide, and so on, eventually producing many cells (only two are shown here).

5. The increased number of B cells produce antibodies, which are part of the antibody-mediated immune system response that eliminates the antigen.

Proliferation of B Cells
Figure 22.15

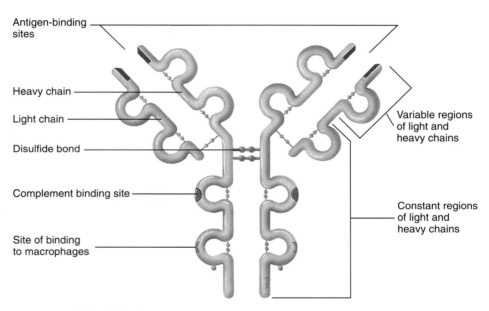

Antigen-binding sites

Heavy chain

Light chain

Disulfide bond

Complement binding site

Site of binding to macrophages

Variable regions of light and heavy chains

Constant regions of light and heavy chains

Structure of an Antibody
Figure 22.16

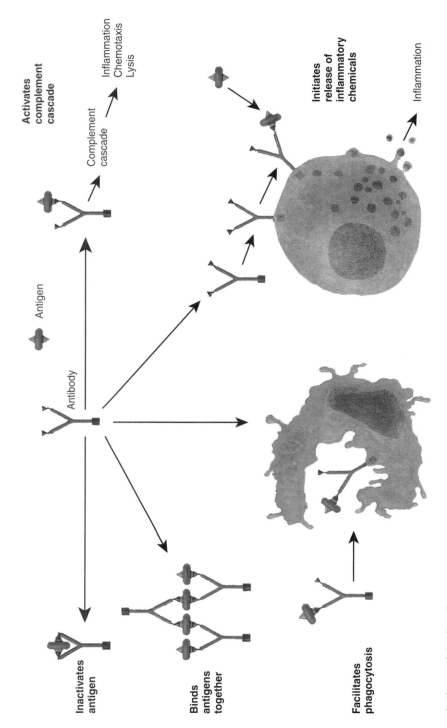

Activates complement cascade

Complement cascade → Inflammation Chemotaxis Lysis

Antigen

Antibody

Inactivates antigen

Binds antigens together

Facilitates phagocytosis

Initiates release of inflammatory chemicals

Inflammation

Actions of Antibodies
Figure 22.17

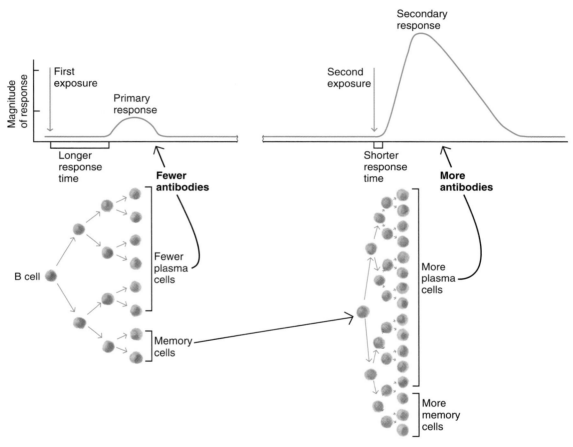

Antibody Production
Figure 22.18

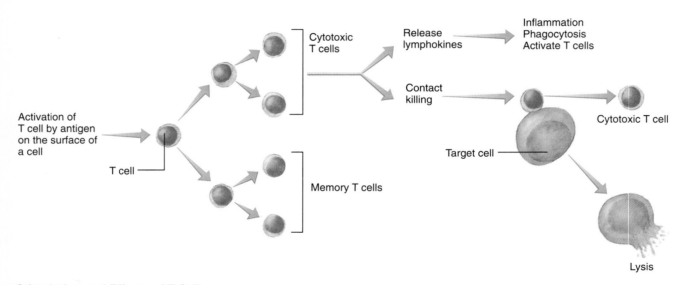

Activation of
T cell by antigen
on the surface of
a cell

T cell

Cytotoxic
T cells

Memory T cells

Release
lymphokines

Inflammation
Phagocytosis
Activate T cells

Contact
killing

Target cell

Cytotoxic T cell

Lysis

Stimulation and Effects of T Cells
Figure 22.19

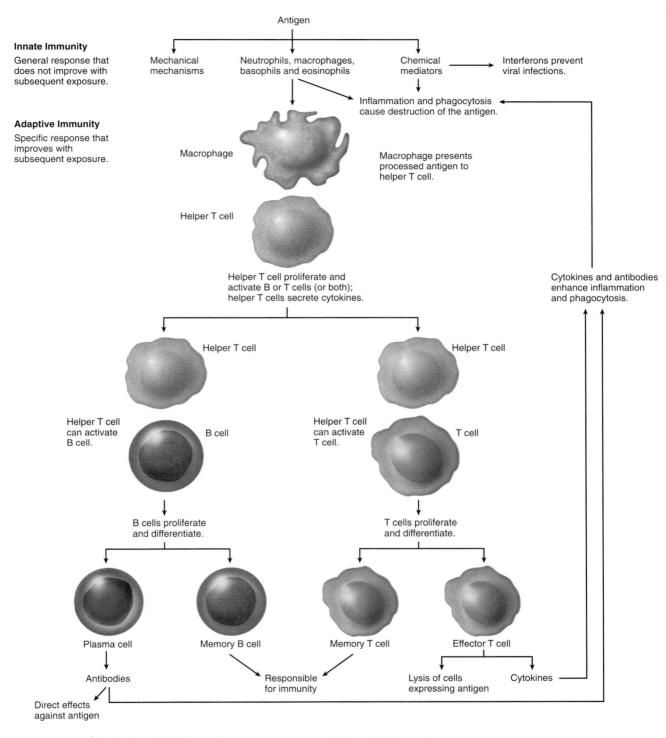

Innate Immunity

General response that does not improve with subsequent exposure.

Adaptive Immunity

Specific response that improves with subsequent exposure.

Antigen

Mechanical mechanisms

Neutrophils, macrophages, basophils and eosinophils

Chemical mediators

Interferons prevent viral infections.

Inflammation and phagocytosis cause destruction of the antigen.

Macrophage

Macrophage presents processed antigen to helper T cell.

Helper T cell

Helper T cell proliferate and activate B or T cells (or both); helper T cells secrete cytokines.

Cytokines and antibodies enhance inflammation and phagocytosis.

Helper T cell

Helper T cell

Helper T cell can activate B cell.

B cell

Helper T cell can activate T cell.

T cell

B cells proliferate and differentiate.

T cells proliferate and differentiate.

Plasma cell

Memory B cell

Memory T cell

Effector T cell

Antibodies

Responsible for immunity

Lysis of cells expressing antigen

Cytokines

Direct effects against antigen

Antibody-mediated immunity

Antibodies act against antigens in solution or on the surfaces of extracellular microorganisms.

Cell-mediated immunity

Effector T cells act against antigens bound to MHC molecules on the surface of cells; effective against viruses and other intracellular microorganisms.

The Major Interactions and Responses of Innate and Adaptive Immunity to an Antigen
Figure 22.20

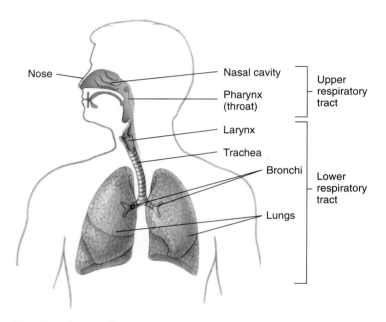

The Respiratory System
Figure 23.1

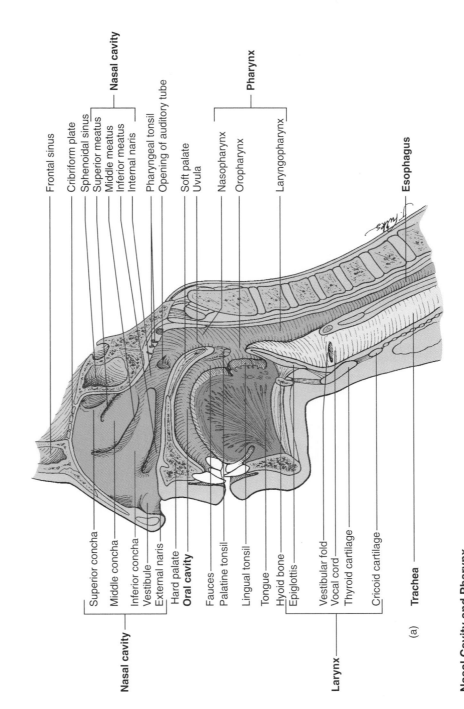

Frontal sinus
Cribriform plate
Sphenoidal sinus
Superior meatus
Middle meatus
Inferior meatus
Internal naris

Nasal cavity

Pharyngeal tonsil
Opening of auditory tube

Soft palate
Uvula

Nasopharynx
Oropharynx

Pharynx

Laryngopharynx

Esophagus

Superior concha
Middle concha
Inferior concha
Vestibule
External naris

Nasal cavity

Hard palate
Oral cavity

Fauces
Palatine tonsil

Lingual tonsil

Tongue
Hyoid bone
Epiglottis

Vestibular fold
Vocal cord
Thyroid cartilage

Cricoid cartilage

Larynx

Trachea

(a)

Nasal Cavity and Pharynx
Figure 23.2a

253

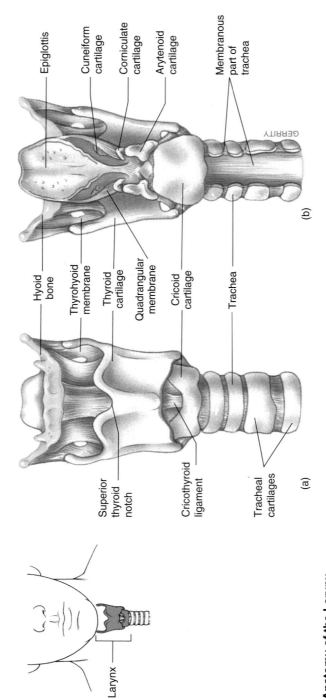

Epiglottis

Cuneiform cartilage

Corniculate cartilage

Arytenoid cartilage

Membranous part of trachea

GERRITY

(b)

Hyoid bone

Thyrohyoid membrane

Thyroid cartilage

Quadrangular membrane

Cricoid cartilage

Trachea

Superior thyroid notch

Cricothyroid ligament

Tracheal cartilages

(a)

Larynx

Anatomy of the Larynx
Figure 23.3a, b

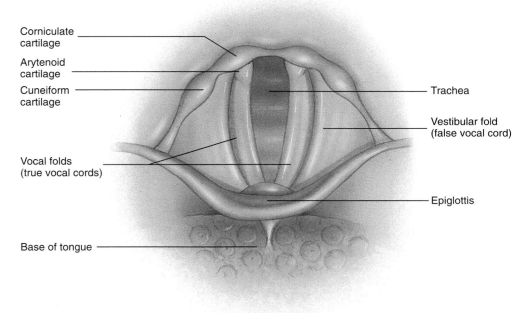

Corniculate cartilage

Arytenoid cartilage

Cuneiform cartilage

Vocal folds (true vocal cords)

Base of tongue

Trachea

Vestibular fold (false vocal cord)

Epiglottis

(a)

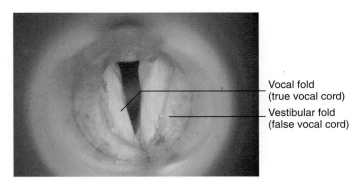

Vocal fold (true vocal cord)

Vestibular fold (false vocal cord)

(b)

Vocal Cords
Figure 23.4

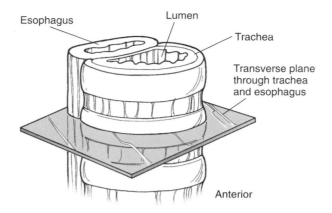

Trachea
Figure 23.5a, icon

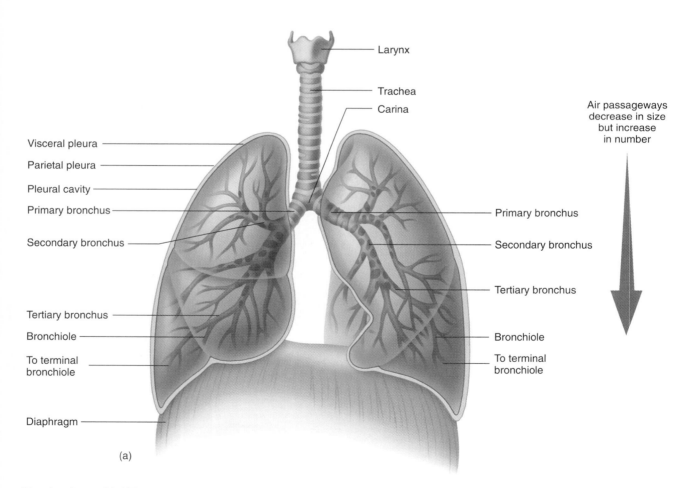

Larynx

Trachea

Carina

Air passageways
decrease in size
but increase
in number

Visceral pleura

Parietal pleura

Pleural cavity

Primary bronchus

Secondary bronchus

Primary bronchus

Secondary bronchus

Tertiary bronchus

Tertiary bronchus

Bronchiole

Bronchiole

To terminal
bronchiole

To terminal
bronchiole

Diaphragm

(a)

Tracheobronchial Tree
Figure 23.6a

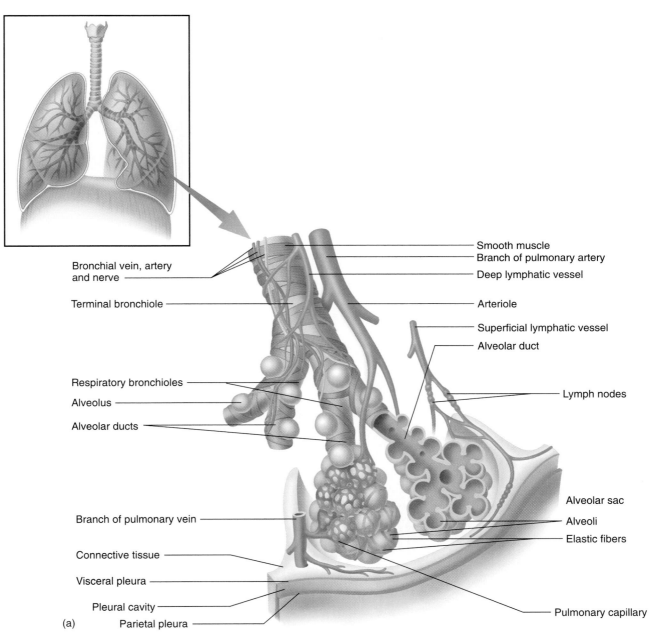

Smooth muscle
Branch of pulmonary artery
Deep lymphatic vessel

Bronchial vein, artery
and nerve

Arteriole

Terminal bronchiole

Superficial lymphatic vessel

Alveolar duct

Respiratory bronchioles

Alveolus

Lymph nodes

Alveolar ducts

Branch of pulmonary vein

Alveolar sac

Alveoli

Connective tissue

Elastic fibers

Visceral pleura

Pleural cavity

(a)

Parietal pleura

Pulmonary capillary

Bronchioles and Alveoli
Figure 23.7a, icon

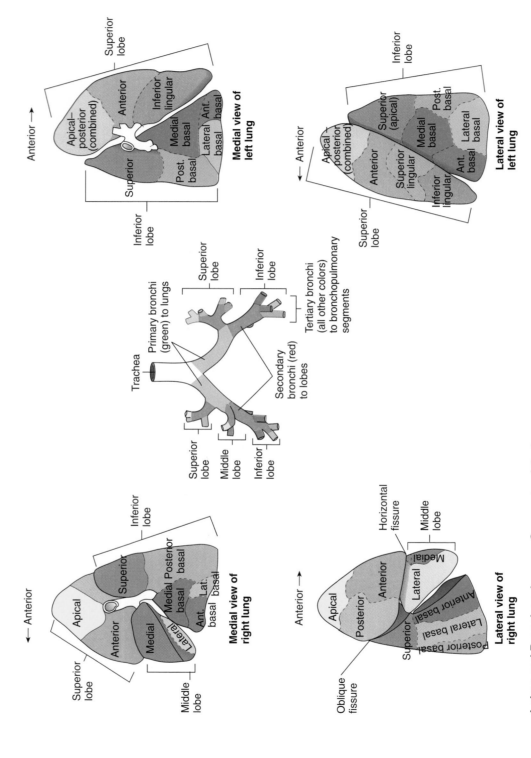

Lobes and Bronchopulmonary Segments of the Lungs
Figure 23.8

Medial view of left lung

Superior lobe

Anterior

Apical–posterior (combined)

Anterior

Inferior lingular

Medial basal

Ant. basal

Lateral basal

Post. basal

Superior

Inferior lobe

Lateral view of left lung

Inferior lobe

Anterior

Apical–posterior (combined)

Superior (apical)

Medial basal

Post. basal

Lateral basal

Anterior

Superior lingular

Inferior lingular

Ant. basal

Superior lobe

Trachea

Primary bronchi (green) to lungs

Superior lobe

Inferior lobe

Tertiary bronchi (all other colors) to bronchopulmonary segments

Secondary bronchi (red) to lobes

Superior lobe

Middle lobe

Inferior lobe

Medial view of right lung

Inferior lobe

Anterior

Apical

Anterior

Superior

Medial

Posterior basal

Medial basal

Lateral

Ant. basal

Lat. basal

Superior lobe

Middle lobe

Lateral view of right lung

Horizontal fissure

Middle lobe

Anterior

Apical

Posterior

Medial

Lateral

Anterior basal

Superior

Lateral basal

Posterior basal

Oblique fissure

258

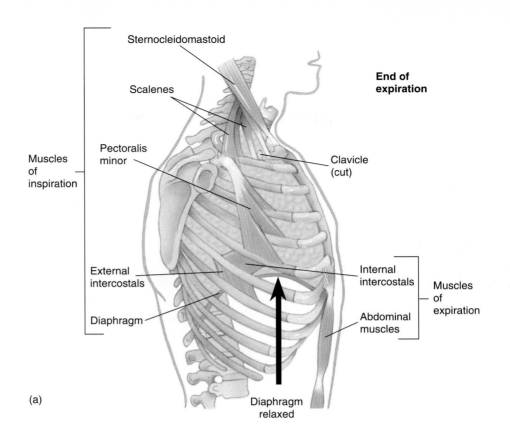

Sternocleidomastoid

Scalenes

Pectoralis minor

Muscles of inspiration

External intercostals

Diaphragm

(a)

End of expiration

Clavicle (cut)

Internal intercostals

Abdominal muscles

Muscles of expiration

Diaphragm relaxed

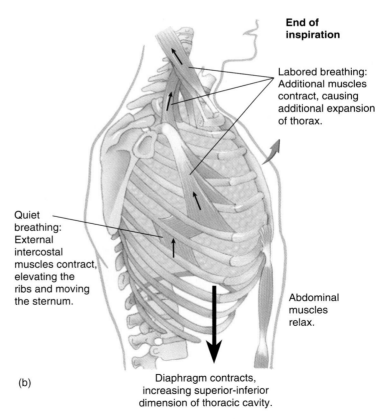

End of inspiration

Labored breathing: Additional muscles contract, causing additional expansion of thorax.

Quiet breathing: External intercostal muscles contract, elevating the ribs and moving the sternum.

Abdominal muscles relax.

(b)

Diaphragm contracts, increasing superior-inferior dimension of thoracic cavity.

Effect of the Muscles of Respiration on Thoracic Volume
Figure 23.9a, b

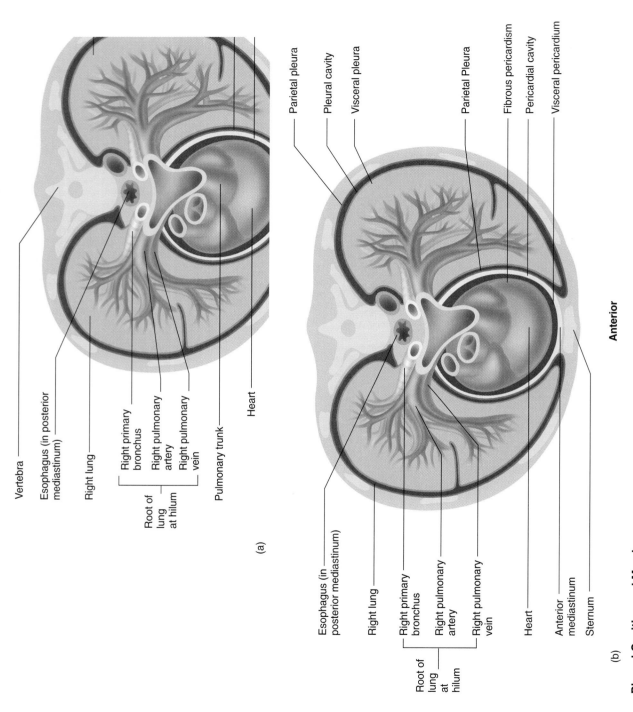

Vertebra

Esophagus (in posterior mediastinum)

Right lung

Root of lung at hilum
{
Right primary bronchus
Right pulmonary artery
Right pulmonary vein
}

Pulmonary trunk

Heart

(a)

Parietal pleura

Pleural cavity

Visceral pleura

Parietal Pleura

Fibrous pericardism

Pericardial cavity

Visceral pericardium

Anterior

Esophagus (in posterior mediastinum)

Right lung

Root of lung at hilum
{
Right primary bronchus
Right pulmonary artery
Right pulmonary vein
}

Heart

Anterior mediastinum

Sternum

(b)

Pleural Cavities and Membranes
Figure 23.10

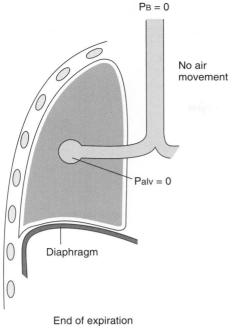

$P_B = 0$

No air
movement

$P_{alv} = 0$

Diaphragm

End of expiration
$P_B = P_{alv}$

(a) Barometric air pressure (P_B) is equal to alveolar pressure (P_{alv}) and there is no air movement.

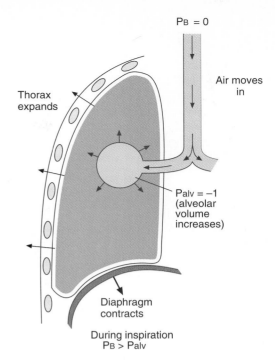

$P_B = 0$

Air moves in

Thorax expands

$P_{alv} = -1$ (alveolar volume increases)

Diaphragm contracts

During inspiration
$P_B > P_{alv}$

(b) Increased thoracic volume results in increased alveolar volume and decreased alveolar pressure. Barometric air presure is greater than alveolar pressure, and air flows into the lungs.

Alveolar Pressure Changes During Inspiration and Expiration
Figure 23.11a, b

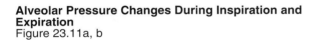

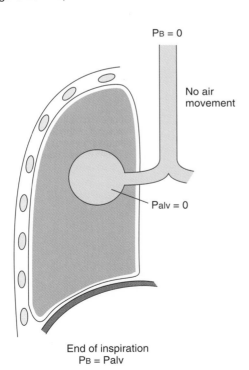

$P_B = 0$

No air
movement

$P_{alv} = 0$

End of inspiration
$P_B = P_{alv}$

(c) Alveolar pressure becomes equal to barometric air pressure, and there is no air movement.

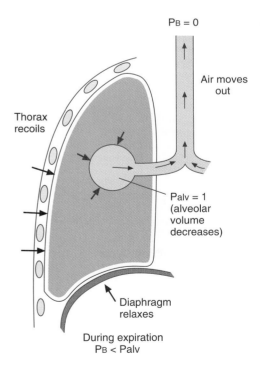

$P_B = 0$

Air moves out

Thorax recoils

$P_{alv} = 1$ (alveolar volume decreases)

Diaphragm relaxes

During expiration
$P_B < P_{alv}$

(d) Decreased thoracic volume results in decreased alveolar volume and increased alveolar pressure. Alveolar pressure is greater than barometric air pressure, and air flows out of the lungs.

Alveolar Pressure Changes During Inspiration and Expiration
Figure 23.11c, d

Changes During Inspiration

Pleural pressure decreases because thoracic volume increases.

As inspiration begins, alveolar pressure decreases below barometric air pressure (0 on the graph) because the decreased pleural pressure causes alveolar volume to increase. By the end of inspiration, alveolar and barometric air pressure are equal.

During inspiration, air flows into the lungs because alveolar pressure is lower than barometric air pressure.

Inspiration   Expiration

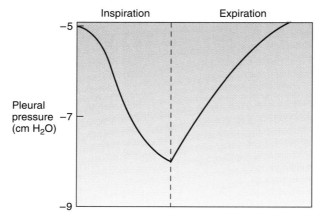

Pleural pressure (cm H₂O)

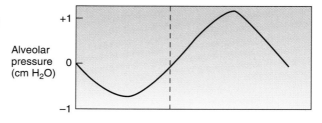

Alveolar pressure (cm H₂O)

Change in lung volume (L)

Time (s)

Changes During Expiration

Pleural pressure increases because thoracic volume decreases.

As expiration begins, alveolar pressure increases above barometric air pressure (0 on the graph) because the increased pleural pressure causes alveolar volume to decrease. By the end of expiration, alveolar and barometric air pressure are equal.

During expiration, air flows out of the lungs because alveolar pressure is greater than barometric air pressure.

Dynamics of a Normal Breathing Cycle
Figure 23.12

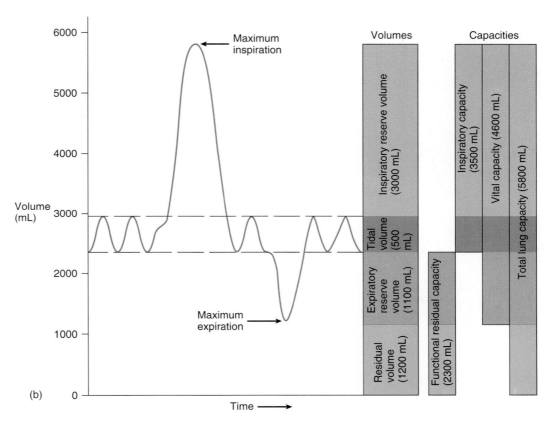

Spirometer, Lung Volumes, and Lung Capacities
Figure 23.13b

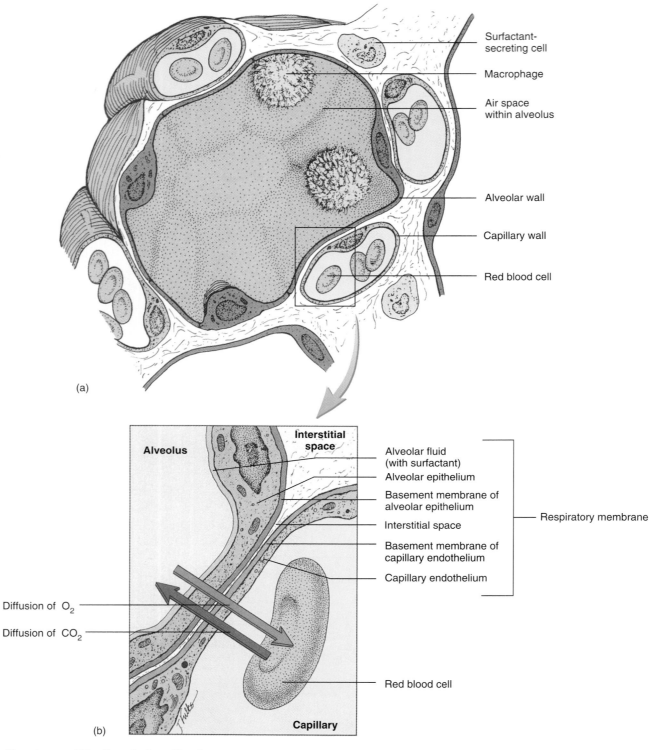

(a)

Surfactant-secreting cell

Macrophage

Air space within alveolus

Alveolar wall

Capillary wall

Red blood cell

Interstitial space

Alveolus

Alveolar fluid (with surfactant)

Alveolar epithelium

Basement membrane of alveolar epithelium

Interstitial space

Basement membrane of capillary endothelium

Capillary endothelium

Respiratory membrane

Diffusion of O_2

Diffusion of CO_2

Red blood cell

Capillary

(b)

Alveolus and the Respiratory Membrane
Figure 23.14b

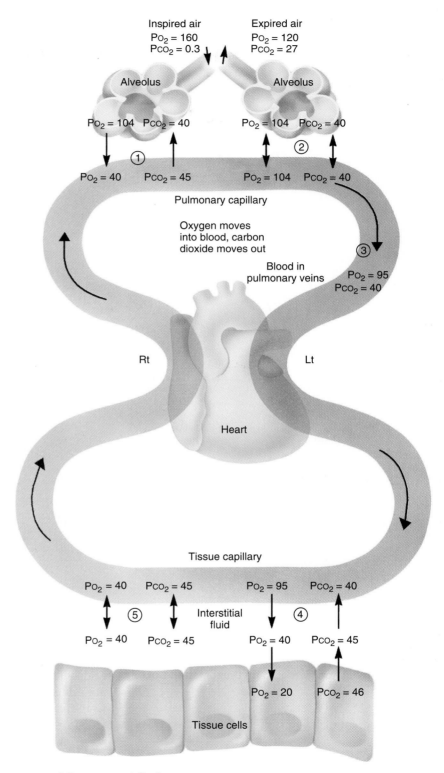

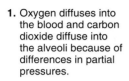

Inspired air
PO₂ = 160
PCO₂ = 0.3

Expired air
PO₂ = 120
PCO₂ = 27

Alveolus

Alveolus

PO₂ = 104   PCO₂ = 40

PO₂ = 104   PCO₂ = 40

①

②

1. Oxygen diffuses into the blood and carbon dioxide diffuse into the alveoli because of differences in partial pressures.

PO₂ = 40   PCO₂ = 45

PO₂ = 104   PCO₂ = 40

Pulmonary capillary

Oxygen moves into blood, carbon dioxide moves out

③

2. As a result of diffusion, the PO₂ in the blood is equal to the PO₂ in the alveoli and the PCO₂ in the blood is equal to the PCO₂ in the alveoli.

Blood in pulmonary veins

PO₂ = 95
PCO₂ = 40

Rt

Lt

3. The PO₂ of blood in the pulmonary veins is less than in the pulmonary capillaries because of mixing with shunted blood.

Heart

4. Oxygen diffuses into the tissue and carbon dioxide diffuse out of the tissue because of differences in partial pressures.

Tissue capillary

PO₂ = 40   PCO₂ = 45

PO₂ = 95   PCO₂ = 40

⑤

④

Interstitial fluid

5. As a result of diffusion, the PO₂ in the blood is equal to the PO₂ in the tissue and the PCO₂ in the blood is equal to the PCO₂ in the tissue. Go back to step 1.

PO₂ = 40   PCO₂ = 45

PO₂ = 40   PCO₂ = 45

PO₂ = 20   PCO₂ = 46

Tissue cells

Changes in the Partial Pressures of Oxygen and Carbon Dioxide
Figure 23.15

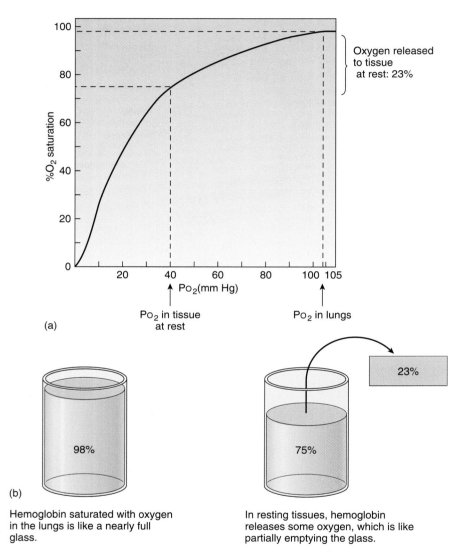

(a)

Oxygen released to tissue at rest: 23%

Po₂ in tissue at rest

Po₂ in lungs

(b)

98%

75%

23%

Hemoglobin saturated with oxygen in the lungs is like a nearly full glass.

In resting tissues, hemoglobin releases some oxygen, which is like partially emptying the glass.

Oxygen-Hemoglobin Dissociation Curve at Rest
Figure 23.16

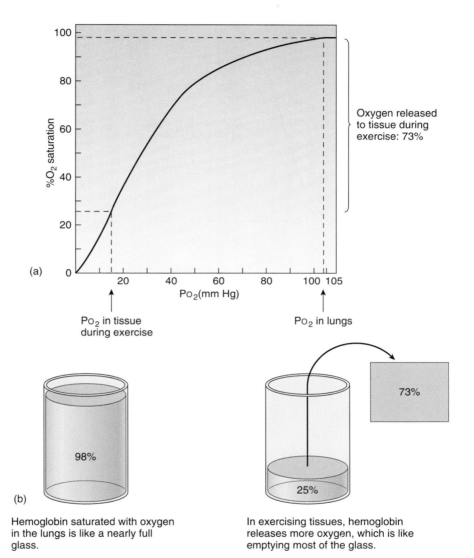

(a)

Oxygen released to tissue during exercise: 73%

Po₂ in tissue during exercise

Po₂ in lungs

(b)

Hemoglobin saturated with oxygen in the lungs is like a nearly full glass.

In exercising tissues, hemoglobin releases more oxygen, which is like emptying most of the glass.

Oxygen-Hemoglobin Dissociation Curve During Exercise
Figure 23.17

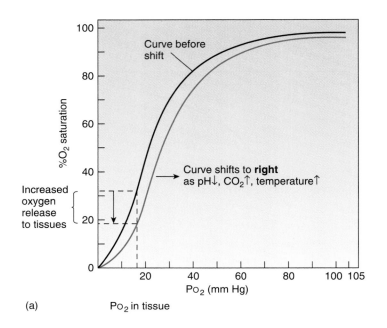

(a)   Po$_2$ in tissue

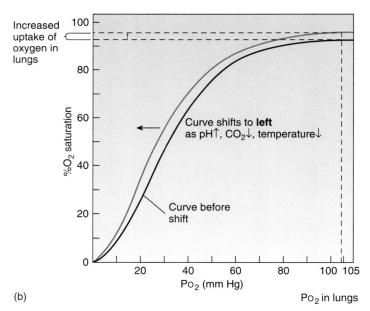

(b)

Po$_2$ in lungs

Effects of Shifting the Oxygen-Hemoglobin Dissociation Curve
Figure 23.18

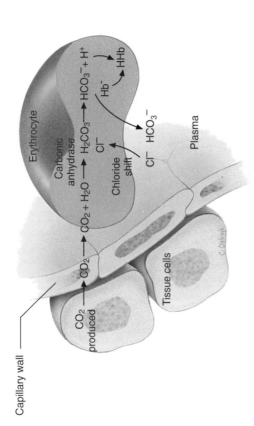

(a) In tissues, carbon dioxide enters erythrocytes and reacts with water to form carbonic acid, which dissociates to form bicarbonate and hydrogen ions. In the chloride shift, bicarbonate ions are exchanged for chloride ions. Hydrogen ions combine with hemoglobin. Lowering the concentration of bicarbonate and hydrogen ions inside erythrocytes promotes the conversion of carbon dioxide to bicarbonate ions.

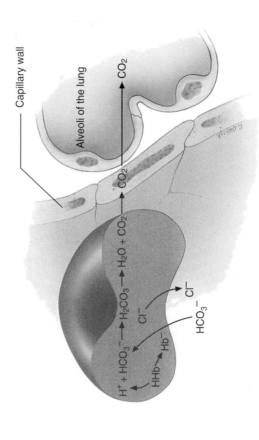

(b) In the lungs, carbon dioxide leaves erythrocytes, resulting in the formation of additional carbon dioxide from carbonic acid. Bicarbonate and hydrogen ions combine to replace the carbonic acid. The bicarbonate ions are exchanged for chloride ions, and the hydrogen ions are released from hemoglobin.

Carbon Dioxide Transport and Chloride Movement
Figure 23.19

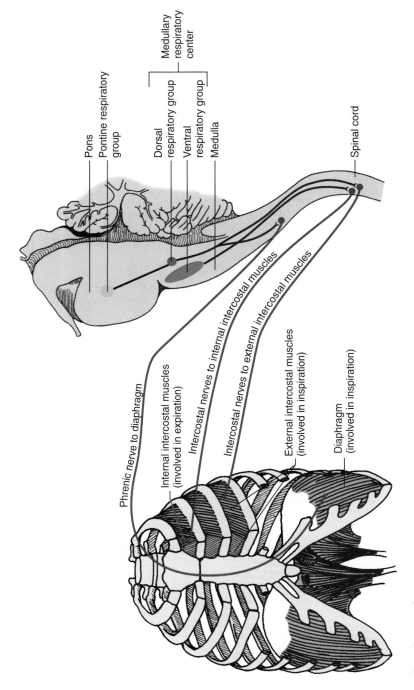

Pons

Pontine respiratory group

Dorsal respiratory group

Ventral respiratory group

Medulla

Medullary respiratory center

Spinal cord

Phrenic nerve to diaphragm

Internal intercostal muscles (involved in expiration)

Intercostal nerves to internal intercostal muscles

Intercostal nerves to external intercostal muscles

External intercostal muscles (involved in inspiration)

Diaphragm (involved in inspiration)

Respiratory Structures in the Brainstem
Figure 23.20

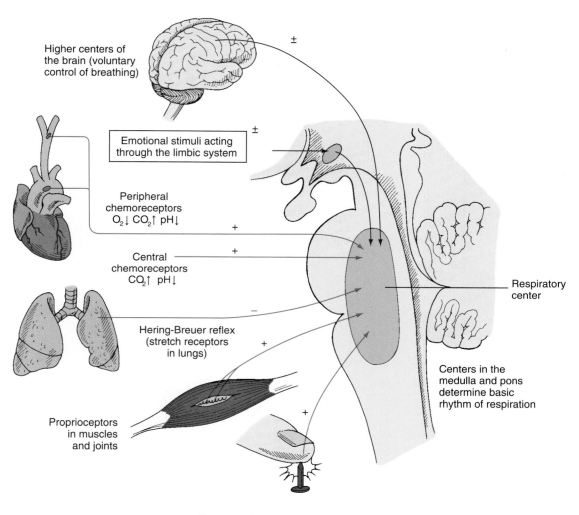

Higher centers of the brain (voluntary control of breathing)

±

Emotional stimuli acting through the limbic system

±

Peripheral chemoreceptors
$O_2\downarrow\ CO_2\uparrow\ pH\downarrow$

+

Central chemoreceptors
$CO_2\uparrow\ pH\downarrow$

+

Hering-Breuer reflex (stretch receptors in lungs)

−

Proprioceptors in muscles and joints

+

Receptors for touch, temperature, and pain stimuli

+

Respiratory center

Centers in the medulla and pons determine basic rhythm of respiration

Modifying Respiration
Figure 23.21

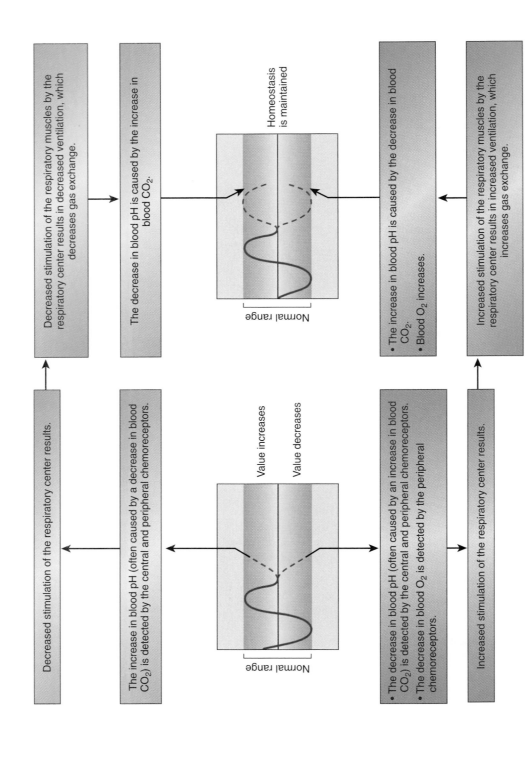

Homeostasis: Regulation of Blood pH and Gases
Figure 23.22

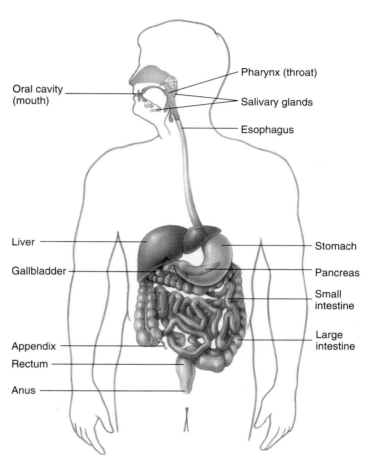

Oral cavity (mouth)

Pharynx (throat)

Salivary glands

Esophagus

Liver

Stomach

Gallbladder

Pancreas

Small intestine

Appendix

Large intestine

Rectum

Anus

Digestive System Depicted in Place in the Body
Figure 24.1

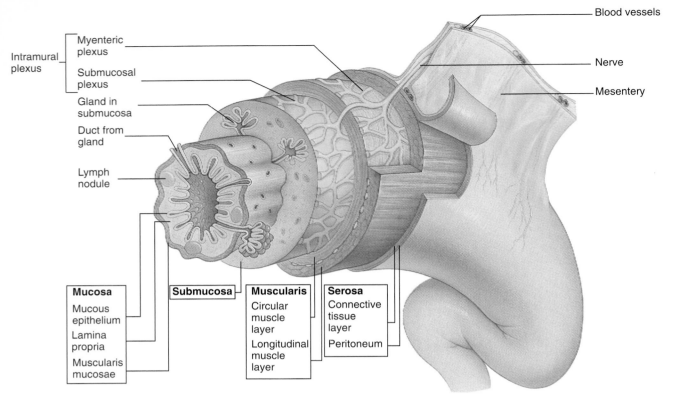

Intramural plexus
- Myenteric plexus
- Submucosal plexus

Gland in submucosa

Duct from gland

Lymph nodule

Blood vessels

Nerve

Mesentery

Mucosa	**Submucosa**	**Muscularis**	**Serosa**
Mucous epithelium		Circular muscle layer	Connective tissue layer
Lamina propria		Longitudinal muscle layer	Peritoneum
Muscularis mucosae			

Digestive Tract Histology
Figure 24.2

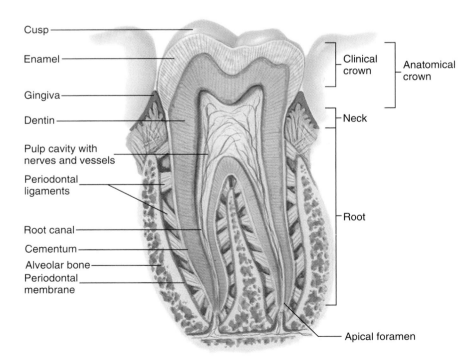

Cusp

Enamel

Gingiva

Dentin

Pulp cavity with nerves and vessels

Periodontal ligaments

Root canal

Cementum

Alveolar bone

Periodontal membrane

Clinical crown

Anatomical crown

Neck

Root

Apical foramen

Molar Tooth in Place in the Alveolar Bone
Figure 24.6

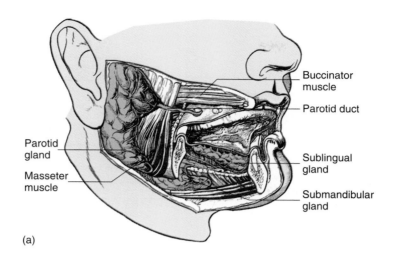

(a)

Buccinator
muscle

Parotid duct

Parotid
gland

Masseter
muscle

Sublingual
gland

Submandibular
gland

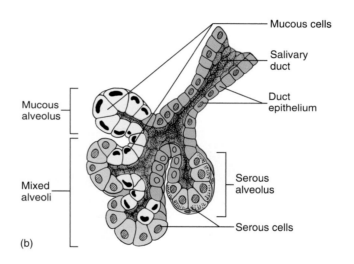

Mucous cells

Salivary
duct

Duct
epithelium

Mucous
alveolus

Mixed
alveoli

Serous
alveolus

Serous cells

(b)

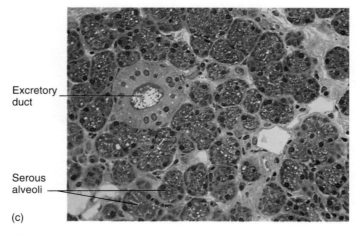

Excretory
duct

Serous
alveoli

(c)

Salivary Glands
Figure 24.7

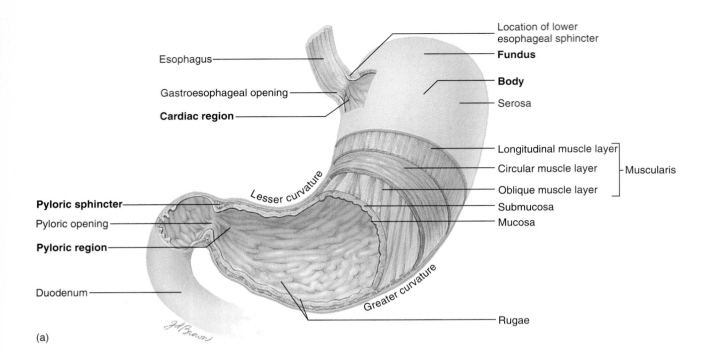

Esophagus

Gastroesophageal opening

Cardiac region

Pyloric sphincter

Pyloric opening

Pyloric region

Duodenum

Lesser curvature

Greater curvature

Location of lower esophageal sphincter

Fundus

Body

Serosa

Longitudinal muscle layer

Circular muscle layer

Oblique muscle layer

Muscularis

Submucosa

Mucosa

Rugae

(a)

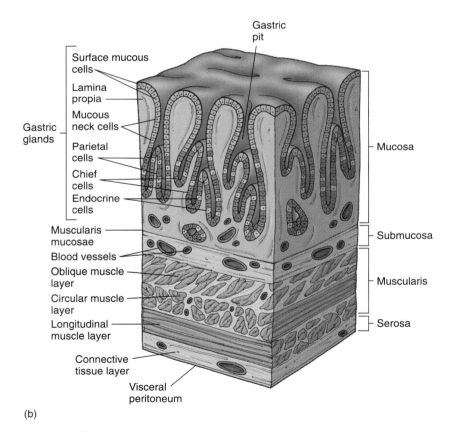

Gastric pit

Gastric glands

Surface mucous cells

Lamina propia

Mucous neck cells

Parietal cells

Chief cells

Endocrine cells

Muscularis mucosae

Blood vessels

Oblique muscle layer

Circular muscle layer

Longitudinal muscle layer

Connective tissue layer

Visceral peritoneum

Mucosa

Submucosa

Muscularis

Serosa

(b)

Anatomy and Histology of the Stomach
Figure 24.8, *(Continued)*

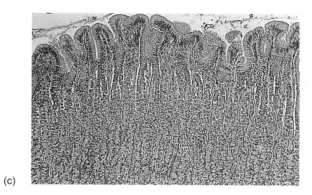

(c)

Figure 24.8, *(Continued)*

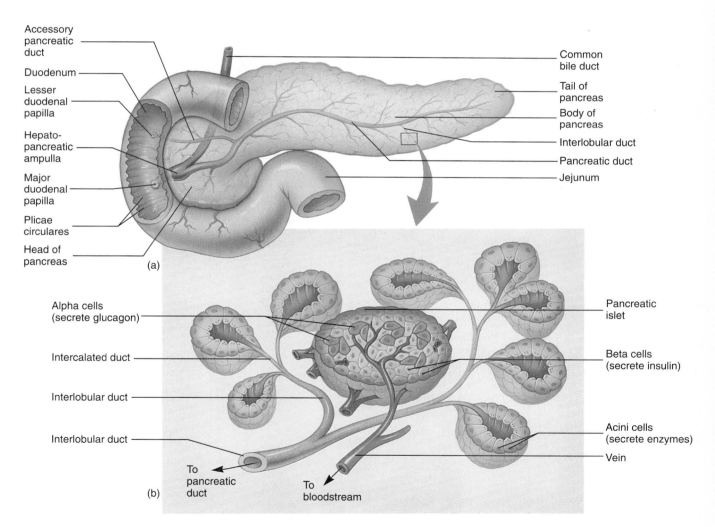

Accessory pancreatic duct

Duodenum

Lesser duodenal papilla

Hepato-pancreatic ampulla

Major duodenal papilla

Plicae circulares

Head of pancreas

Common bile duct

Tail of pancreas

Body of pancreas

Interlobular duct

Pancreatic duct

Jejunum

(a)

Alpha cells (secrete glucagon)

Intercalated duct

Interlobular duct

Interlobular duct

To pancreatic duct

To bloodstream

Pancreatic islet

Beta cells (secrete insulin)

Acini cells (secrete enzymes)

Vein

(b)

Anatomy and Histology of the Duodenum and Pancreas
Figure 24.10

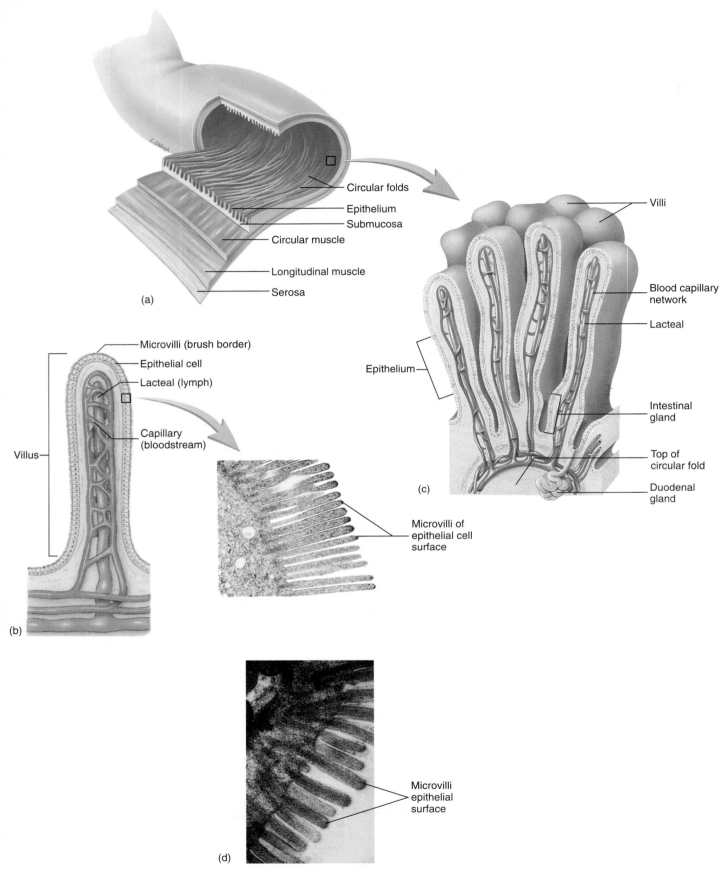

(a)

Circular folds
Epithelium
Submucosa
Circular muscle
Longitudinal muscle
Serosa

Villi
Blood capillary network
Lacteal
Epithelium
Intestinal gland
Top of circular fold
Duodenal gland

(c)

Microvilli (brush border)
Epithelial cell
Lacteal (lymph)
Capillary (bloodstream)
Villus

(b)

Microvilli of epithelial cell surface

Microvilli epithelial surface

(d)

Anatomy and Histology of the Duodenum
Figure 24.11

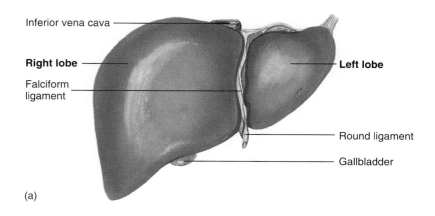

Inferior vena cava

Right lobe

Falciform ligament

Left lobe

Round ligament

Gallbladder

(a)

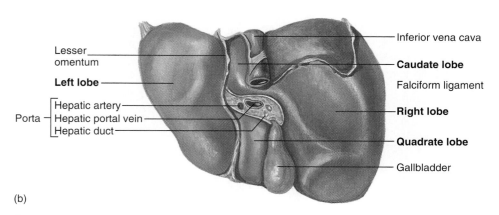

Lesser omentum

Left lobe

Porta { Hepatic artery
Hepatic portal vein
Hepatic duct }

Inferior vena cava

Caudate lobe

Falciform ligament

Right lobe

Quadrate lobe

Gallbladder

(b)

Anatomy and Histology of the Liver
Figure 24.12, *(Continued)*

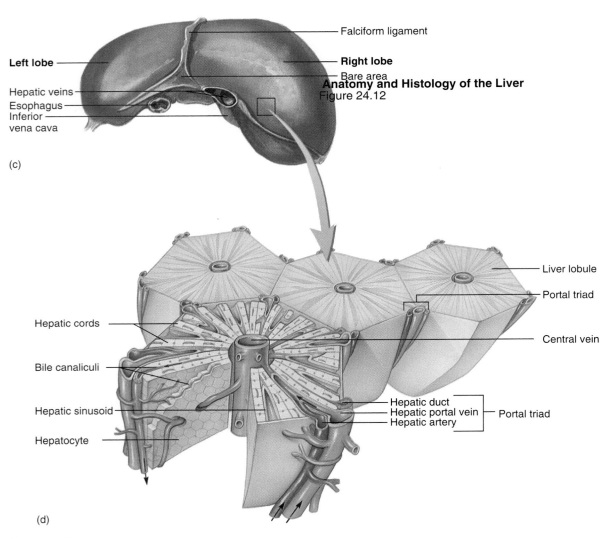

Falciform ligament

Left lobe

Right lobe

Bare area

Hepatic veins

Esophagus

Inferior
vena cava

Anatomy and Histology of the Liver
Figure 24.12

(c)

Liver lobule

Portal triad

Hepatic cords

Central vein

Bile canaliculi

Hepatic sinusoid

Hepatic duct
Hepatic portal vein ⎫ Portal triad
Hepatic artery

Hepatocyte

(d)

Figure 24.12

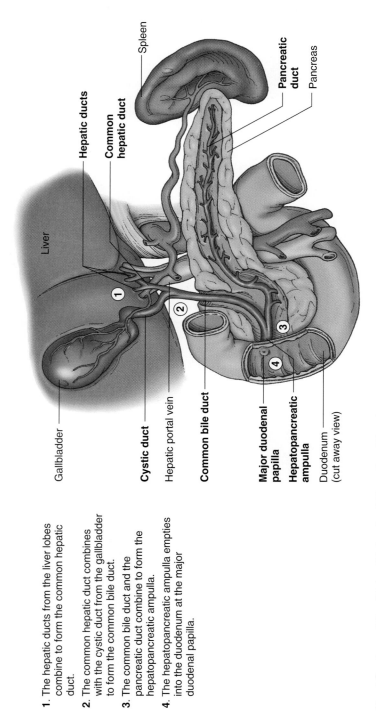

Liver

Gallbladder

Hepatic ducts

Common
hepatic duct

Spleen

Pancreatic
duct

Pancreas

Cystic duct

Hepatic portal vein

Common bile duct

**Major duodenal
papilla**

**Hepatopancreatic
ampulla**

Duodenum
(cut away view)

1. The hepatic ducts from the liver lobes combine to form the common hepatic duct.

2. The common hepatic duct combines with the cystic duct from the gallbladder to form the common bile duct.

3. The common bile duct and the pancreatic duct combine to form the hepatopancreatic ampulla.

4. The hepatopancreatic ampulla empties into the duodenum at the major duodenal papilla.

Duct System of the Major Abdominal Digestive Glands
Figure 24.13

281

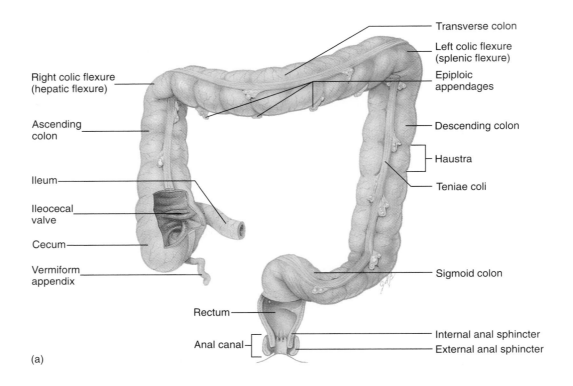

Transverse colon

Left colic flexure
(splenic flexure)

Right colic flexure
(hepatic flexure)

Epiploic
appendages

Ascending
colon

Descending colon

Haustra

Ileum

Teniae coli

Ileocecal
valve

Cecum

Vermiform
appendix

Sigmoid colon

Rectum

Internal anal sphincter

Anal canal

External anal sphincter

(a)

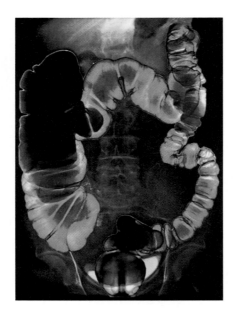

(b)

Large Intestine and Anal Canal
Figure 24.14

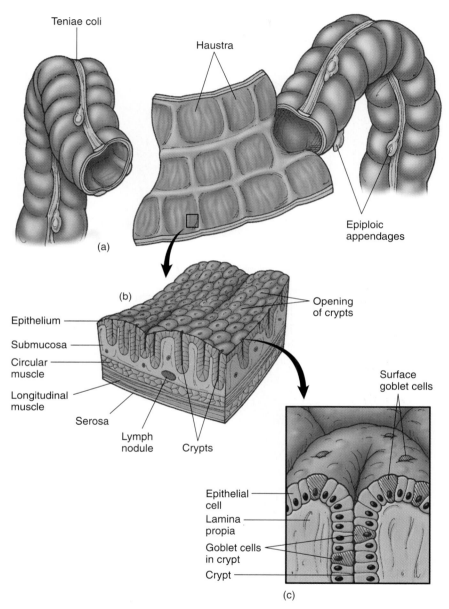

Teniae coli

Haustra

Epiploic
appendages

(a)

(b)

Epithelium

Submucosa

Circular
muscle

Longitudinal
muscle

Serosa

Lymph
nodule

Crypts

Opening
of crypts

Surface
goblet cells

Epithelial
cell

Lamina
propia

Goblet cells
in crypt

Crypt

(c)

Histology of the Large Intestine
Figure 24.15

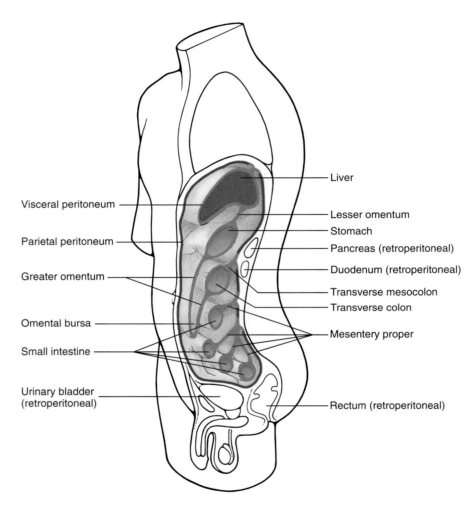

Visceral peritoneum

Parietal peritoneum

Greater omentum

Omental bursa

Small intestine

Urinary bladder
(retroperitoneal)

Liver

Lesser omentum

Stomach

Pancreas (retroperitoneal)

Duodenum (retroperitoneal)

Transverse mesocolon

Transverse colon

Mesentery proper

Rectum (retroperitoneal)

Peritoneum and Mesenteries
Figure 24.16

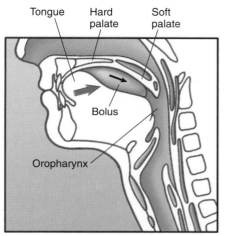

Tongue Hard palate Soft palate

Bolus

Oropharynx

(a) During the voluntary phase, a bolus of food (*yellow*) is pushed by the tongue against the hard palate and posteriorly toward the oropharynx (*red arrow* indicates tongue movement; *black arrow* indicates movement of the bolus).

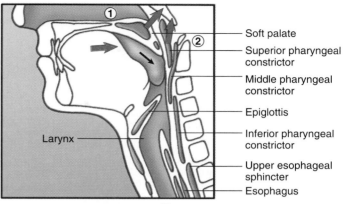

Soft palate

Superior pharyngeal constrictor

Middle pharyngeal constrictor

Epiglottis

Inferior pharyngeal constrictor

Upper esophageal sphincter

Esophagus

Larynx

(b) During the pharyngeal phase, ① the soft palate is elevated, closing off the nasopharynx. ② The pharynx is elevated (*red arrows* indicate muscle movement).

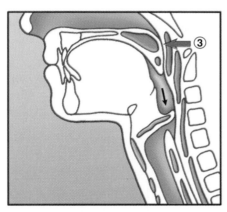

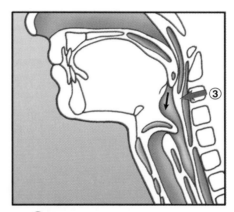

(c) Successive constriction of the pharyngeal constrictors ③ from superior to inferior (*red arrows*) forces the bolus through the pharynx and into the esophagus. As this occurs, the epiglottis is bent down over the opening of the larynx largely by the force of the bolus pressing against it.

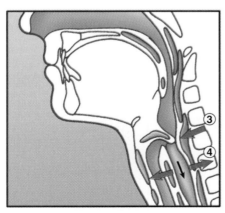

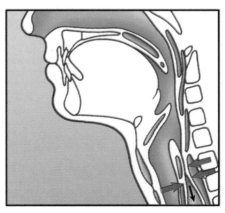

(d) The ④ upper esophageal sphincter relaxes (*outwardly directed red arrows*), allowing the bolus to enter the esophagus.

(e) During the esophageal phase, the bolus is moved by peristaltic contractions of the esophagus toward the stomach (*inwardly directed red arrows*).

Three Phases of Swallowing (Deglutition)
Figure 24.17

1. The taste or smell of food, tactile sensations of food in the mouth, or even thoughts of food stimulate the vagal nuclei in the medulla oblongata *(blue arrow)*.
2. Parasympathetic sensations are carried by the vagus nerves to the stomach *(green arrow)*.
3. Preganglionic parasympathetic vagus nerve fibers synapse with postganglionic neurons in the myenteric plexus of the stomach.
4. Postganglionic fibers directly stimulate secretion of parietal and chief cells and stimulate gastrin secretion by endocrine cells.
5. Gastrin is carried through the circulation back to other parts of the stomach *(pink arrow)*, where it also stimulates secretion by parietal and chief cells.

(a)

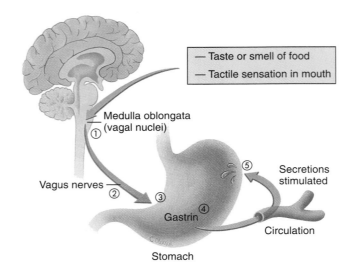

— Taste or smell of food
— Tactile sensation in mouth

Medulla oblongata
(vagal nuclei)
①

Vagus nerves
②

③

Gastrin ④

⑤ Secretions stimulated

Circulation

Stomach

Three Phases of Gastric Secretion (Cephalic Stage)
Figure 24.19a

Efferent vagal impulses *(green arrow)* and local impulses stimulate secretions in the parietal, chief, and mucous cells. The low pH in the stomach can inhibit these secretions.

Distention of the stomach with food stimulates local reflexes in the stomach *(pink arrow)* and vagus nerve impulses to the medulla *(blue arrow)*.

(b)

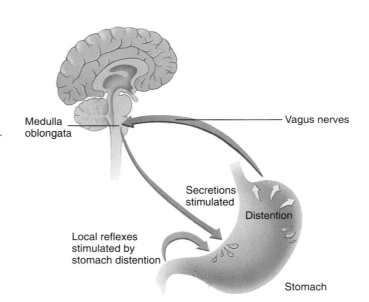

Medulla oblongata

Vagus nerves

Secretions stimulated

Distention

Local reflexes stimulated by stomach distention

Stomach

Three Phases of Gastric Secretion (Gastric Phase)
Figure 24.19b

1. The presence in the duodenum of chyme with a pH greater than 3 *(blue arrow)* or containing amino acids and peptides stimulates gastric secretions through vagus nerve pathways *(blue and green arrows)* and through secretion of gastrin *(pink arrows to the right)*.
2. The presence in the duodenum of hypotonic chyme with a pH less than 2 *(red arrow)* or chyme containing fat digestion products inhibits gastric secretions by three mechanisms (3–5).
3. Afferent vagal impulses *(lower purple arrow)* inhibit efferent impulses from the vagal nuclei of the medulla oblongata *(upper purple arrow)*.
4. Secretin inhibits gastrin secretion in the duodenum *(orange arrow)*.
5. Secretin, gastric inhibitory polypeptide, and cholecystokinin produced by the duodenum *(brown arrows)* inhibit gastric secretions in the stomach.

(c)

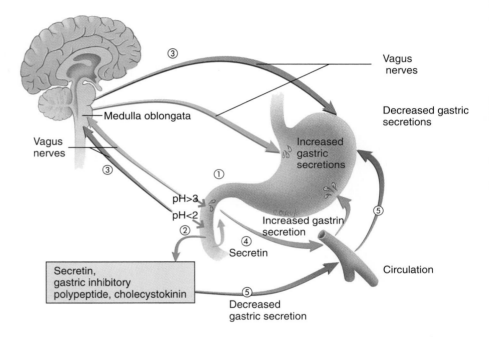

Three Phases of Gastric Secretion (Intestinal Phase)
Figure 24.19c

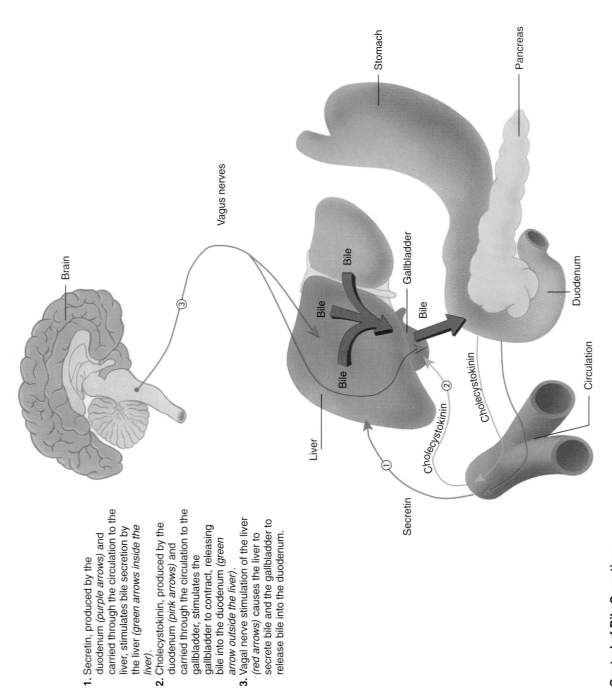

1. Secretin, produced by the duodenum *(purple arrows)* and carried through the circulation to the liver, stimulates bile secretion by the liver *(green arrows inside the liver)*.

2. Cholecystokinin, produced by the duodenum *(pink arrows)* and carried through the circulation to the gallbladder, stimulates the gallbladder to contract, releasing bile into the duodenum *(green arrow outside the liver)*.

3. Vagal nerve stimulation of the liver *(red arrows)* causes the liver to secrete bile and the gallbladder to release bile into the duodenum.

Brain

Vagus nerves

Stomach

Pancreas

Bile

Bile

Bile

Bile

Gallbladder

Liver

Duodenum

Cholecystokinin

Cholecystokinin

Circulation

Secretin

Control of Bile Secretion
Figure 24.21

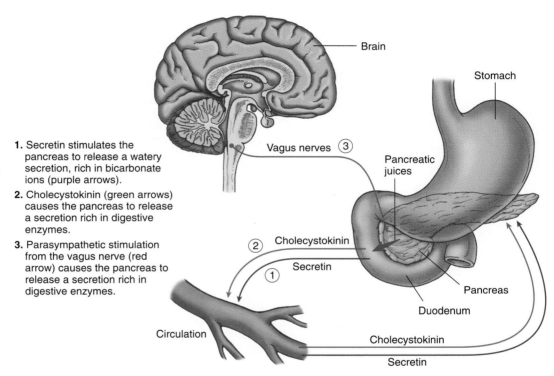

1. Secretin stimulates the pancreas to release a watery secretion, rich in bicarbonate ions (purple arrows).

2. Cholecystokinin (green arrows) causes the pancreas to release a secretion rich in digestive enzymes.

3. Parasympathetic stimulation from the vagus nerve (red arrow) causes the pancreas to release a secretion rich in digestive enzymes.

Brain

Stomach

Vagus nerves ③

Pancreatic juices

Cholecystokinin

② Cholecystokinin

① Secretin

Secretin

Pancreas

Duodenum

Circulation

Cholecystokinin

Secretin

Control of Pancreatic Secretion
Figure 24.23

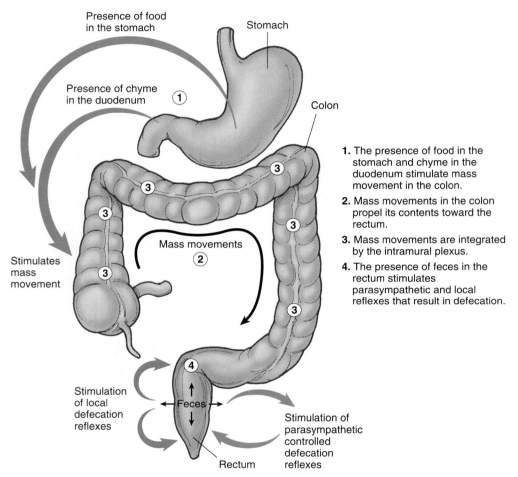

Presence of food
in the stomach

Stomach

Presence of chyme
in the duodenum ①

Colon

1. The presence of food in the
stomach and chyme in the
duodenum stimulate mass
movement in the colon.

2. Mass movements in the colon
propel its contents toward the
rectum.

3. Mass movements are integrated
by the intramural plexus.

4. The presence of feces in the
rectum stimulates
parasympathetic and local
reflexes that result in defecation.

Mass movements ②

Stimulates
mass
movement

Stimulation
of local
defecation
reflexes

Feces

Stimulation of
parasympathetic
controlled
defecation
reflexes

Rectum

Reflexes in the Colon and Rectum
Figure 24.24

1. Cotransport carrier molecules in the cell membrane of the brush border bind sodium ions (Na$^+$) and glucose or galactose.
2. Sodium ions pass into the cell, down their concentration gradient, by facilitated diffusion.
3. Sodium ions are actively transported, in exchange for potassium (K$^+$) ions, out of the basilar side of the cell, maintaining the sodium concentration gradient inside the cell, which is responsible for cotransport.
4. The electrochemical energy produced by the movement of sodium ions down their concentration gradient is used to transport glucose or galactose into the cell against their concentration gradients.
5. Glucose or galactose move out of the epithelial cell by facilitated diffusion.

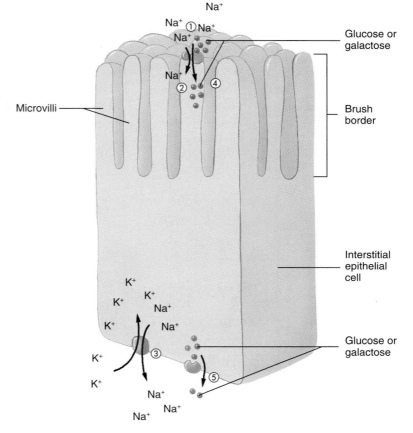

Glucose and Galactose Absorption
Figure 24.25

1. Lipids *(yellow spheres)* associate with bile salts *(blue spheres with rods attached)* to form a micelle.
2. In a micelle the hydrophobic ends of the bile salts are directed inward toward the lipid core, and the hydrophilic ends are directed outward toward the water environment of the digestive tract.
3. When a micelle contacts the microvilli of an epithelial cell, the bile salts are released *(also see inset)* and some of the lipids diffuse directly into the cell. Triacylglycerol is broken down into fatty acids and glycerol, which are taken up by the epithelial cells.
4. In the cell, fatty acids and glycerol combine to form triacylglycerol, which is packaged by the addition of proteins *(green spheres in inset)* into chylomicrons. Cholesterol and phospholipids are also packaged in chylomicrons.
5. The chylomicrons enter the lacteals of the villi and pass into the lymphatic system.

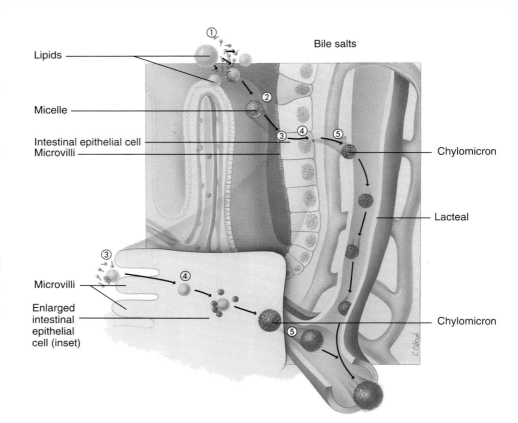

Lipid Absorption
Figure 24.26

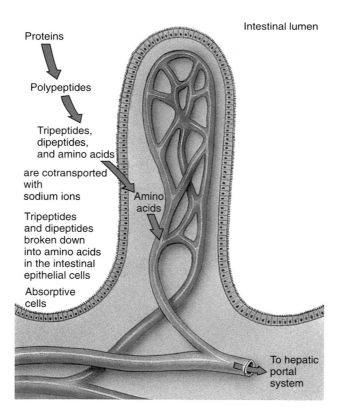

Mechanisms of Amino Acid Absorption
Figure 24.30

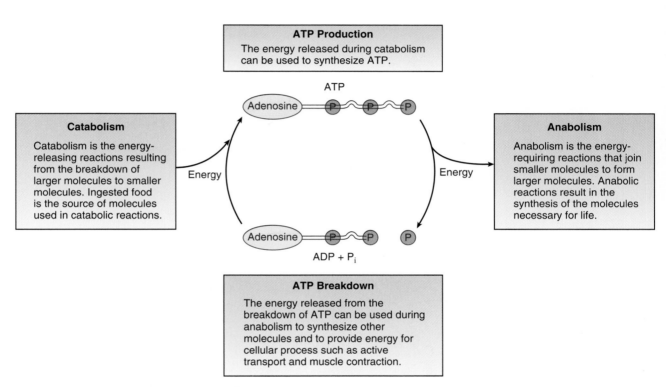

ATP Production
The energy released during catabolism can be used to synthesize ATP.

ATP

Adenosine ━━ P ∿ P ∿ P

Catabolism

Catabolism is the energy-releasing reactions resulting from the breakdown of larger molecules to smaller molecules. Ingested food is the source of molecules used in catabolic reactions.

Energy

Energy

Anabolism

Anabolism is the energy-requiring reactions that join smaller molecules to form larger molecules. Anabolic reactions result in the synthesis of the molecules necessary for life.

Adenosine ━━ P ∿ P   P

$ADP + P_i$

ATP Breakdown

The energy released from the breakdown of ATP can be used during anabolism to synthesize other molecules and to provide energy for cellular process such as active transport and muscle contraction.

ATP Coupling of Catabolic and Anabolic Reactions
Figure 25.2

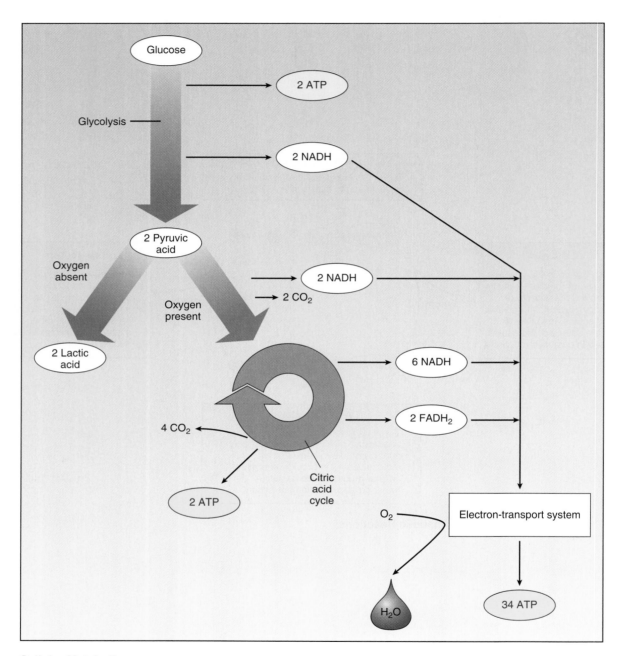

Cellular Metabolism
Figure 25.3

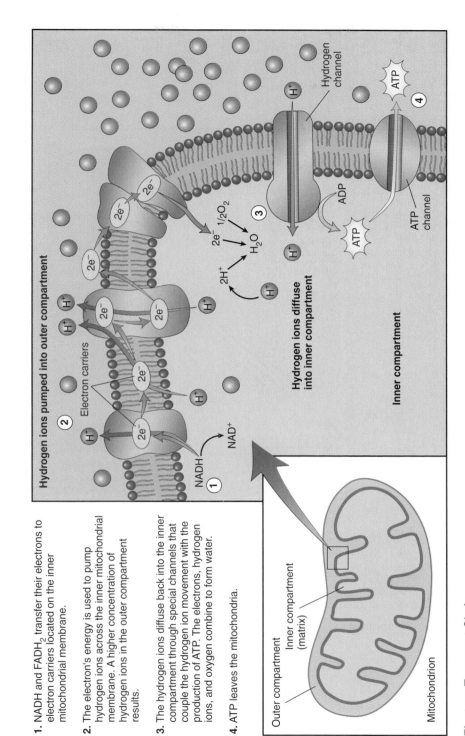

Hydrogen ions pumped into outer compartment

② Electron carriers

① NADH → NAD⁺

$2H^+$

$2e^-\ 1/2O_2$ → H_2O

③ **Hydrogen ions diffuse into inner compartment**

Hydrogen channel

ATP channel

ADP → ATP

④ ATP

Inner compartment

Outer compartment

Inner compartment (matrix)

Mitochondrion

1. NADH and FADH₂ transfer their electrons to electron carriers located on the inner mitochondrial membrane.

2. The electron's energy is used to pump hydrogen ions across the inner mitochondrial membrane. A higher concentration of hydrogen ions in the outer compartment results.

3. The hydrogen ions diffuse back into the inner compartment through special channels that couple the hydrogen ion movement with the production of ATP. The electrons, hydrogen ions, and oxygen combine to form water.

4. ATP leaves the mitochondria.

Electron Transport Chain
Figure 25.7

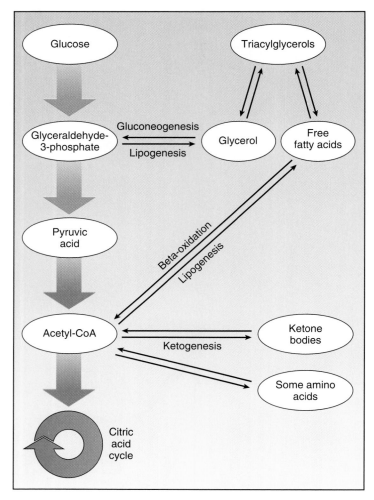

Lipid Metabolism
Figure 25.8

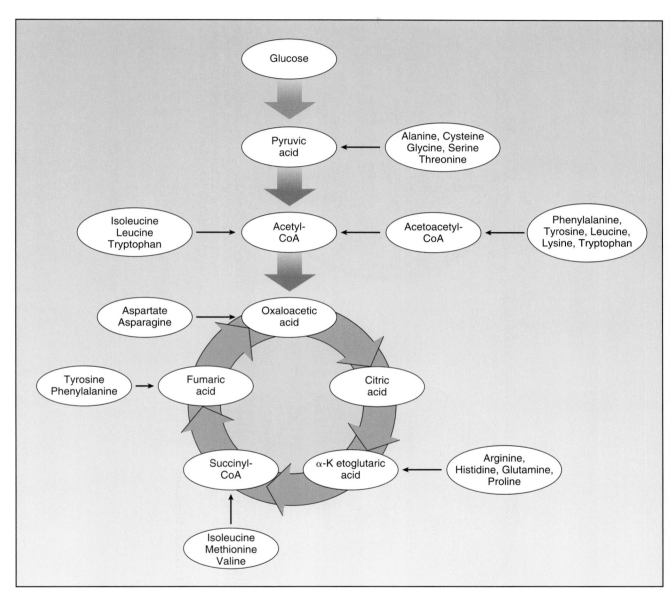

Amino Acid Metabolism
Figure 25.9

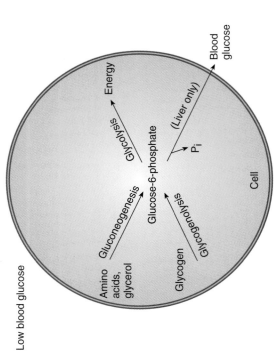

Low blood glucose

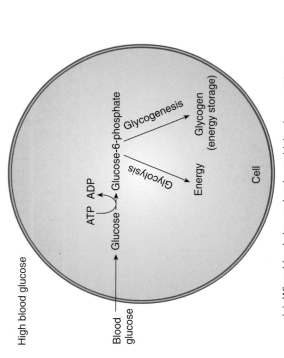

High blood glucose

(a) When blood glucose levels are high, glucose enters the cell and is phosphorylated to form glucose-6-phosphate, which can enter glycolysis or glycogenesis.

(b) When blood glucose levels drop, glucose-6-phosphate can be produced through glycogenolysis or gluconeogenesis. Glucose-6-phosphate can enter glycolysis, or the phosphate group can be removed in liver tissue, and glucose released into the blood.

Interconversion of Nutrient Molecules
Figure 25.12

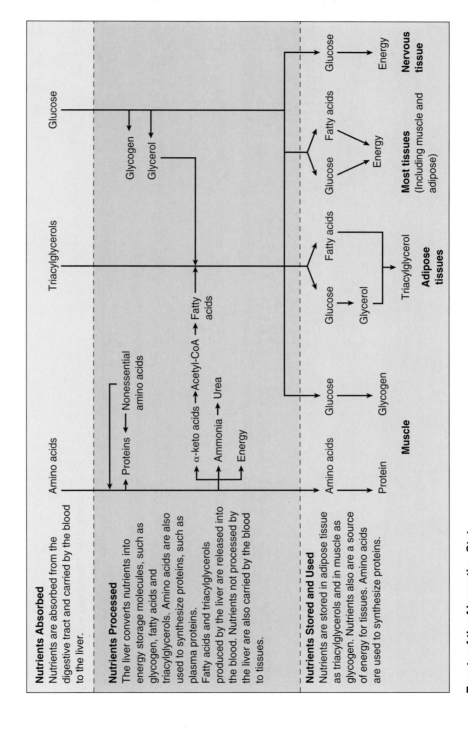

Events of the Absorptive State
Figure 25.13

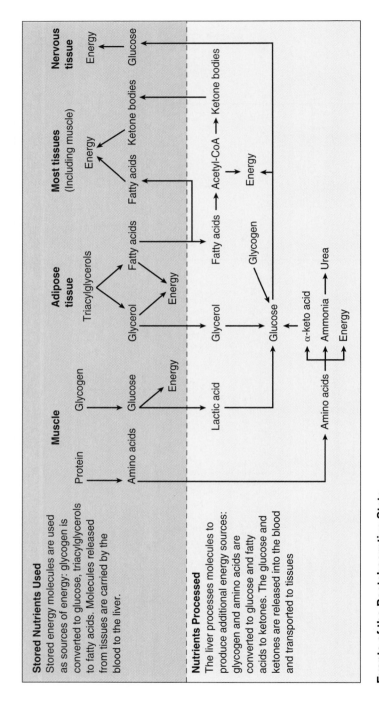

Events of the Postabsorptive State
Figure 25.14

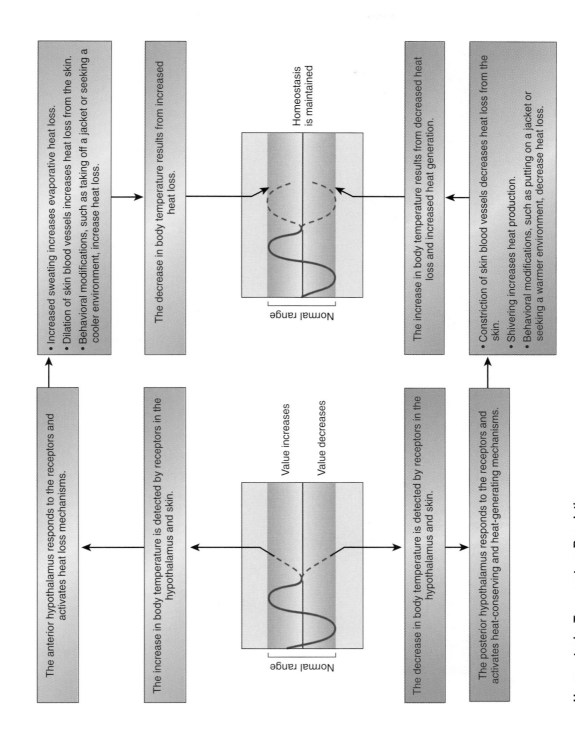

Homeostasis: Temperature Regulation
Figure 25.16

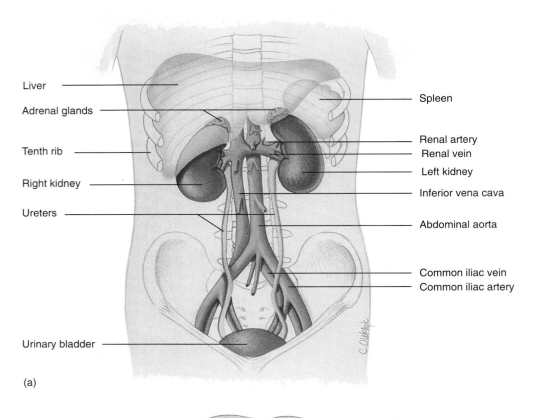

Liver

Adrenal glands

Tenth rib

Right kidney

Ureters

Urinary bladder

Spleen

Renal artery
Renal vein
Left kidney
Inferior vena cava

Abdominal aorta

Common iliac vein
Common iliac artery

(a)

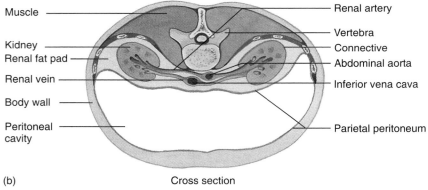

Muscle

Kidney
Renal fat pad

Renal vein

Body wall

Peritoneal
cavity

Renal artery

Vertebra
Connective
Abdominal aorta

Inferior vena cava

Parietal peritoneum

(b)   Cross section

Anatomy of the Urinary System
Figure 26.1

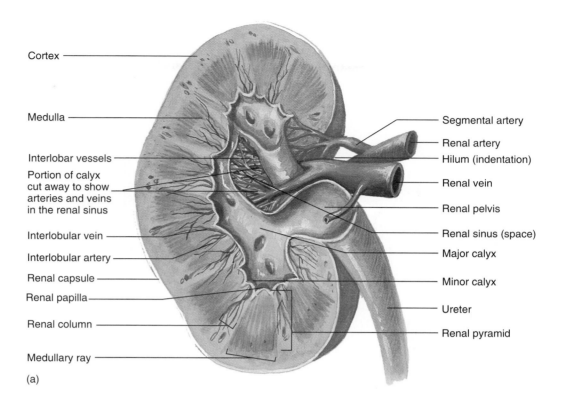

Cortex

Medulla

Interlobar vessels

Portion of calyx cut away to show arteries and veins in the renal sinus

Interlobular vein

Interlobular artery

Renal capsule

Renal papilla

Renal column

Medullary ray

Segmental artery

Renal artery

Hilum (indentation)

Renal vein

Renal pelvis

Renal sinus (space)

Major calyx

Minor calyx

Ureter

Renal pyramid

(a)

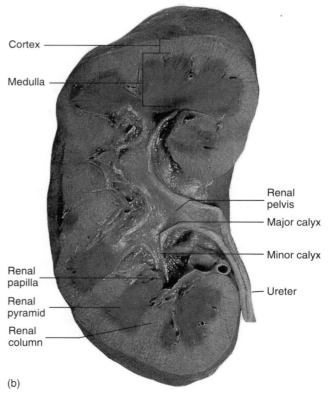

Cortex

Medulla

Renal pelvis

Major calyx

Minor calyx

Renal papilla

Ureter

Renal pyramid

Renal column

(b)

A Longitudinal Section of the Kidney and Ureter
Figure 26.2a, b

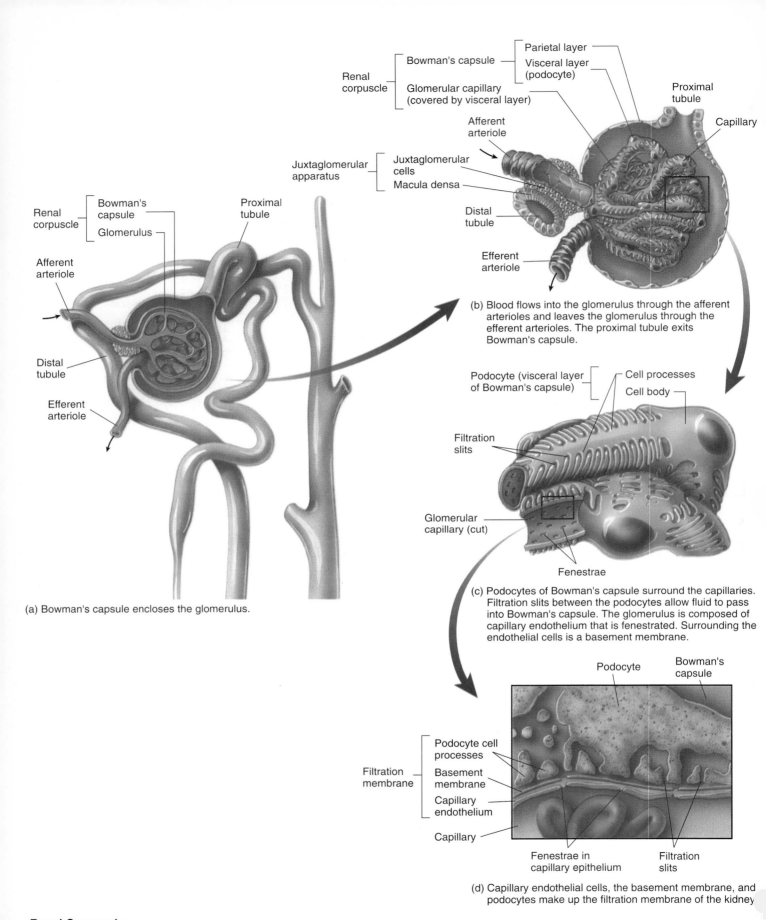

Parietal layer

Visceral layer (podocyte)

Renal corpuscle
Bowman's capsule

Glomerular capillary (covered by visceral layer)

Proximal tubule

Afferent arteriole

Capillary

Juxtaglomerular apparatus
Juxtaglomerular cells

Macula densa

Distal tubule

Efferent arteriole

(b) Blood flows into the glomerulus through the afferent arterioles and leaves the glomerulus through the efferent arterioles. The proximal tubule exits Bowman's capsule.

Renal corpuscle
Bowman's capsule

Glomerulus

Proximal tubule

Afferent arteriole

Distal tubule

Efferent arteriole

(a) Bowman's capsule encloses the glomerulus.

Podocyte (visceral layer of Bowman's capsule)
Cell processes

Cell body

Filtration slits

Glomerular capillary (cut)

Fenestrae

(c) Podocytes of Bowman's capsule surround the capillaries. Filtration slits between the podocytes allow fluid to pass into Bowman's capsule. The glomerulus is composed of capillary endothelium that is fenestrated. Surrounding the endothelial cells is a basement membrane.

Podocyte

Bowman's capsule

Filtration membrane
Podocyte cell processes

Basement membrane

Capillary endothelium

Capillary

Fenestrae in capillary epithelium

Filtration slits

(d) Capillary endothelial cells, the basement membrane, and podocytes make up the filtration membrane of the kidney.

Renal Corpuscle
Figure 26.4

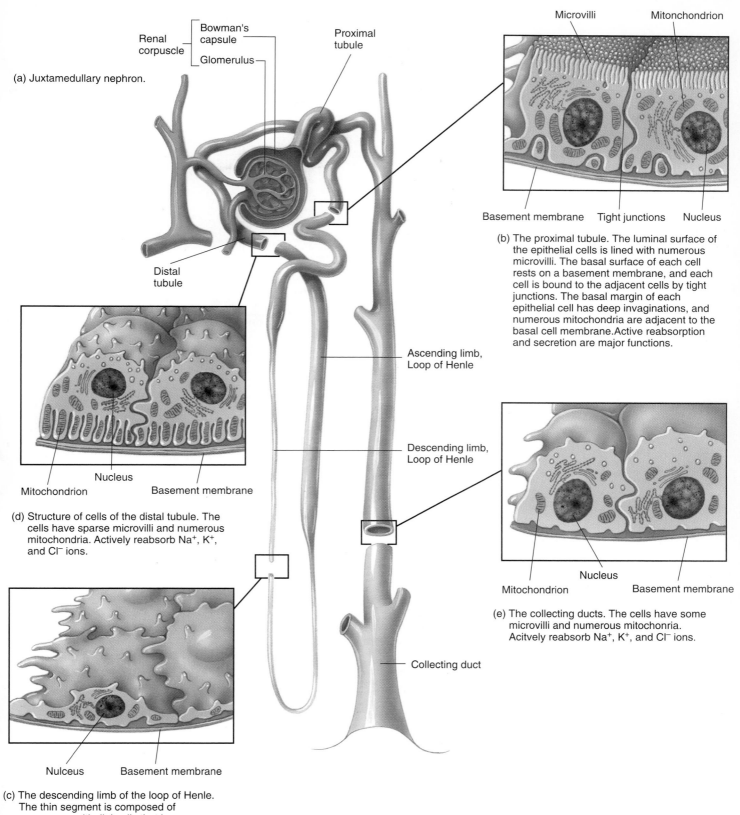

(a) Juxtamedullary nephron.

Renal corpuscle
Bowman's capsule
Glomerulus

Proximal tubule

Distal tubule

Microvilli   Mitonchondrion

Basement membrane   Tight junctions   Nucleus

(b) The proximal tubule. The luminal surface of the epithelial cells is lined with numerous microvilli. The basal surface of each cell rests on a basement membrane, and each cell is bound to the adjacent cells by tight junctions. The basal margin of each epithelial cell has deep invaginations, and numerous mitochondria are adjacent to the basal cell membrane. Active reabsorption and secretion are major functions.

Ascending limb, Loop of Henle

Descending limb, Loop of Henle

Nucleus
Mitochondrion   Basement membrane

(d) Structure of cells of the distal tubule. The cells have sparse microvilli and numerous mitochondria. Actively reabsorb Na$^+$, K$^+$, and Cl$^-$ ions.

Nucleus
Mitochondrion   Basement membrane

(e) The collecting ducts. The cells have some microvilli and numerous mitochonria. Acitvely reabsorb Na$^+$, K$^+$, and Cl$^-$ ions.

Collecting duct

Nulceus   Basement membrane

(c) The descending limb of the loop of Henle. The thin segment is composed of squamous epithelial cells that have microvilli and contain a relatively small number of mitochondria. Water easily diffuses from the thin segment into the interstitial space.

Histology of the Nephron
Figure 26.5

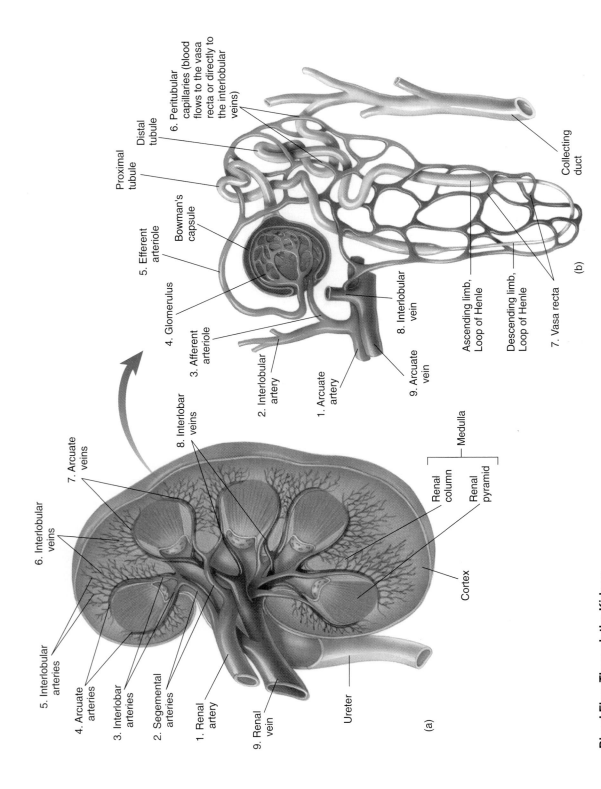

Blood Flow Through the Kidney
Figure 26.6

5. Interlobular arteries

6. Interlobular veins

4. Arcuate arteries

7. Arcuate veins

3. Interlobar arteries

2. Segmental arteries

1. Renal artery

9. Renal vein

Ureter

(a)

8. Interlobar veins

Renal column

Renal pyramid

Medulla

Cortex

7. Arcuate veins

8. Interlobular veins

9. Arcuate vein

1. Arcuate artery

2. Interlobular artery

3. Afferent arteriole

4. Glomerulus

5. Efferent arteriole

Bowman's capsule

Proximal tubule

Distal tubule

6. Peritubular capillaries (blood flows to the vasa recta or directly to the interlobular veins)

Ascending limb, Loop of Henle

Descending limb, Loop of Henle

7. Vasa recta

Collecting duct

(b)

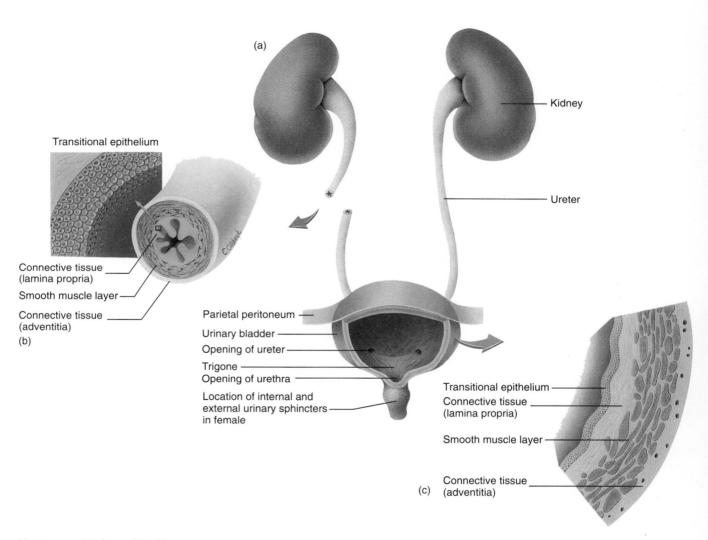

(a)

Kidney

Ureter

Transitional epithelium

Connective tissue
(lamina propria)

Smooth muscle layer

Connective tissue
(adventitia)

(b)

Parietal peritoneum

Urinary bladder

Opening of ureter

Trigone

Opening of urethra

Location of internal and
external urinary sphincters
in female

Transitional epithelium

Connective tissue
(lamina propria)

Smooth muscle layer

Connective tissue
(adventitia)

(c)

Ureters and Urinary Bladder
Figure 26.7

Urine formation results from the following three processes:

1. Filtration   a. Filtration is the movement of materials
                      across the filtration membrane into the
                      lumen of Bowman's capsule to form filtrate.

2. Reabsorption   b. Solutes are reabsorbed across the wall of
                      the nephron by transport processes, such
                      as active transport and cotransport.

                  c. Water is reabsorbed across the wall of the
                      nephron by osmosis.

3. Secretion   d. Solutes are secreted across the wall of the
                      nephron into the filtrate.

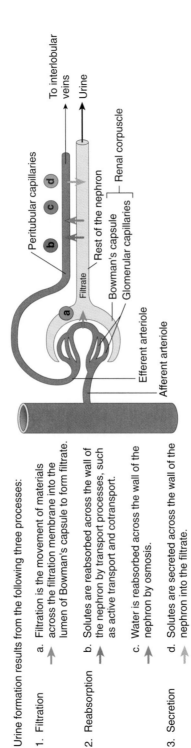

Urine Formation
Figure 26.8

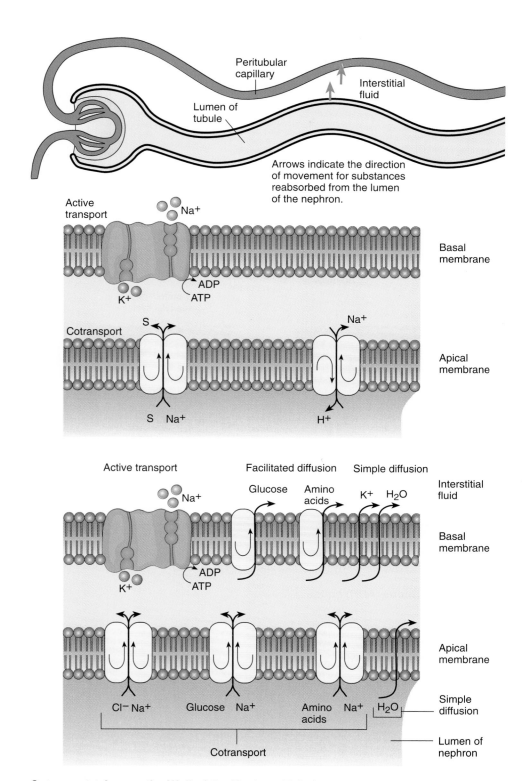

Cotransport Across the Wall of the Nephron Tubule
Figure 26.10

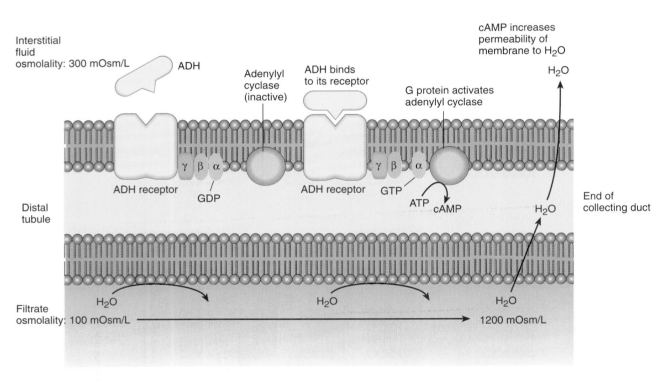

The Effect of Antidiuretic Hormone (ADH) on the Nephron
Figure 26.11

There is an increase in the osmolality of the interstitial fluid from 300 mOsm/L in the cortex to 1200 mOsm/L in the medulla of the kidney. The vasa recta, transport of solutes from the ascending limb of Henle's loop, and recycling of urea maintain the high concentration of solutes in the interstitial fluid of the medulla of the kidney.

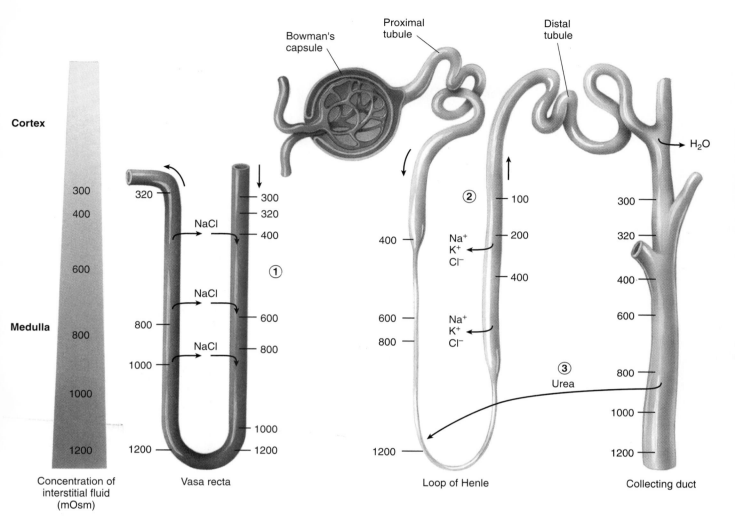

1. The vasa recta function as a countercurrent system. Blood flows from the cortex into the medulla and then from the medulla to the cortex. As it flows to the medulla, water diffuses from the relatively dilute blood into the interstitial fluid, and solutes diffuse from the concentrated interstitial fluid into the blood. At the apex of the medulla the blood is the same concentration as the interstitial fluid. As the concentrated blood flows from the medulla toward the cortex, solutes diffuse out of the vasa recta into the interstitial fluid and water diffuses from the interstitial fluid into the blood. By the time blood leaves the vasa recta, the concentration and volume of blood are only slightly larger than when the blood entered the vasa recta. Thus, blood is supplied to the medulla, excess water and electrolytes are carried away, and the high concentration of the medullary interstitial fluid is maintained.

2. Water diffuses out of the descending limb of Henle's loop into the interstitial space. Na⁺, K⁺, and Cl⁻ are transported out ot the ascending limb of Henle's loop by active transport into the interstitial fluid.

3. There is a high concentration of urea in the medulla of the kidney. The descending limb of Henle's loop and the collecting duct are permeable to urea. All other areas of the nephron are relatively impermeable to urea. Urea diffuses into the descending limb of Henle's loop, passes through the ascending limb and distal tubule, and then diffuses out of the collecting duct into the medullary interstitial fluid. Thus, urea cycles from the interstitial fluid, to the descending limb of Henle's loop, and from the collecting duct into the interstitial fluid of the medulla again.

Maintenance of Medullary Interstitial Concentration
Figure 26.12

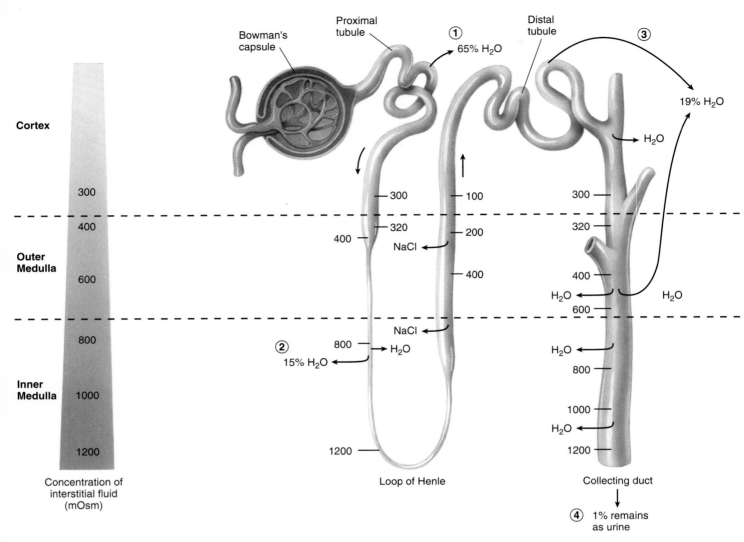

Cortex

Outer
Medulla

Inner
Medulla

300
400
600
800
1000
1200

Concentration of
interstitial fluid
(mOsm)

Bowman's
capsule

Proximal
tubule

① 65% H₂O

Distal
tubule

③

19% H₂O

H₂O

300 — 100
400 — 320 — 200
NaCl
400

300 —
320 —

400 —
H₂O —
600 —

H₂O

NaCl
② 800 — H₂O
15% H₂O

H₂O —
800 —

1000 —
H₂O —
1200 —
Collecting duct

1200 — Loop of Henle

④ 1% remains
as urine

Urine Concentrating Mechanism
Figure 26.13

1. Approximately 180 L of filtrate enters the nephrons each day; however, 65% of that volume is reabsorbed in the proximal tubule. In the proximal tubule, solute molecules are transported from the lumen of the tubule into the interstitial fluid. Water follows the reabsorbed solutes because the cells of the tubule wall are permeable to water.

2. Approximatley 15% of the filtrate volume is reabsorbed in the descending limb of Henle's loop. The descending limb of Henle's loop passes through the concentrated fluid of the medulla. Because the wall of the descending limb of Henle's loop is permeable to water, water diffuses from the tubule into the more concentrated interstitial fluid. By the time the filtrate reaches the apex of the medulla of the kidney, the concentration of the filtrate is equal to the concentration of the the interstitial fluid.

The ascending limb of Henle's loop is not permeable to water, and Na⁺, K⁺, and Cl⁻ ions are actively transported from the filtrate into the interstitial fluid. Consequently, the volume of the filtrate does not change as it passes through the interstitial fluid, but the concentration is greatly reduced. By the time the filtrate reaches the cortex of the kidney, the concentration is approximately 100 mOsm/L, which is less concentrated than the interstitial fluid of the cortex (300 mOsm/L).

3. The distal nephron and collecting duct are permeable to water if ADH is present. If ADH is present, water diffuses from the more concentrated filtrate into the interstitial fluid. By the time the filtrate has reached the apex of the medulla an additional 19% of the filtrate has been reabsorbed, and

4. 1% or less remains as urine.

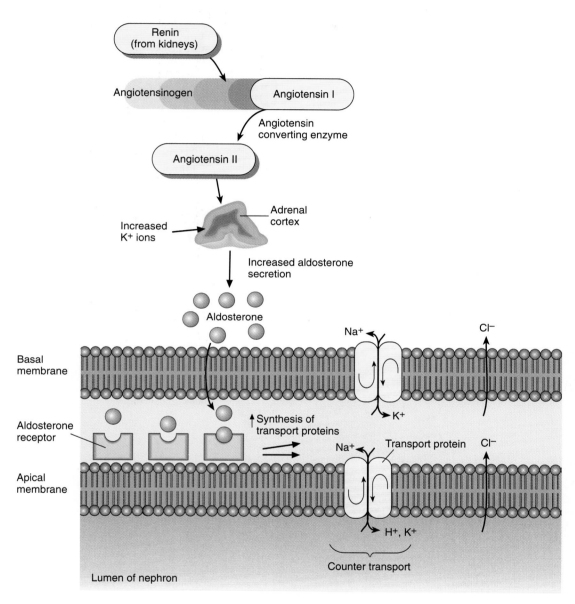

Effect of Aldosterone on the Distal Convoluted Tubule
Figure 26.14

Control of the micturation reflex by higher brain centers

A. Ascending pathways carry an increased frequency of action potentials up the spinal cord to the brain when the urinary bladder becomes stretched.

B. Descending pathways carry action potentials to the sacral region of the spinal cord to inhibit the micturation reflex tonically and to stimulate the reflex when stretch of the urinary bladder produces the conscious urge to urinate and when one voluntarily chooses to urinate.

Micturation reflex

1. Urine in the urinary bladder stretches the bladder wall.

2. Action potentials produced by the stretch receptors (blue line) are carried along pelvic nerves to the sacral region of the spinal cord.

3. Action potentials are carried by the parasympathetic nerves (black line) to relax the internal urinary sphincter and to contract the smooth muscles of the urinary bladder. Decreased action potentials carried by the somatic motor nerves (green line) cause the external urinary sphincter to relax.

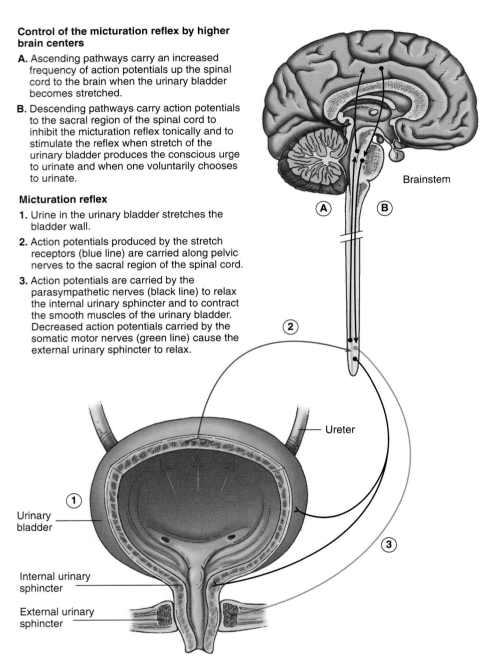

Control of the Micturition Reflex
Figure 26.17

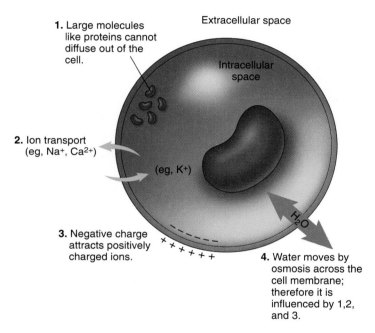

1. Large molecules like proteins cannot diffuse out of the cell.

Extracellular space

Intracellular space

2. Ion transport (eg, Na^+, Ca^{2+})

(eg, K^+)

H_2O

3. Negative charge attracts positively charged ions.

4. Water moves by osmosis across the cell membrane; therefore it is influenced by 1,2, and 3.

Cellular Distribution of Water and Solutes
Figure 27.1

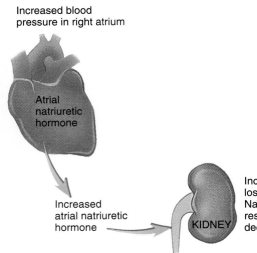

Increased blood pressure in right atrium

Atrial natriuretic hormone

Increased atrial natriuretic hormone → KIDNEY

Increased water loss and increased Na⁺ ion excretion results in decreased BP.

Increased blood pressure (BP) in the right atrium of the heart causes increased secretion of atrial natriuretic hormone, which increases sodium ion excretion and water loss in the form of urine.

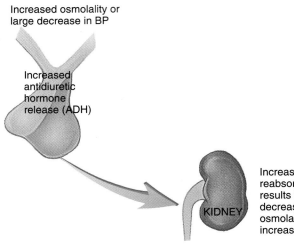

Increased osmolality or large decrease in BP

Increased antidiuretic hormone release (ADH)

KIDNEY

Increased water reabsorption results in decreased osmolality and increased BP.

Increased blood osmolality affects hypothalamic neurons, and decreased blood pressure affects baroreceptors in the aortic arch, carotid sinuses, and atrium. As a result of these stimuli, an increased rate of ADH secretion from the posterior pituitary results, which increases water reabsorption.

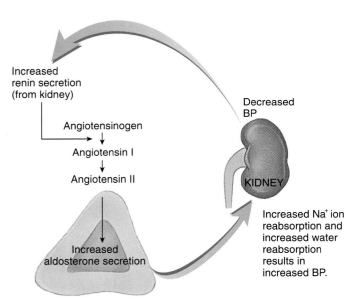

Increased renin secretion (from kidney)

Angiotensinogen
↓
Angiotensin I
↓
Angiotensin II
↓
Increased aldosterone secretion

Decreased BP

KIDNEY

Increased Na⁺ ion reabsorption and increased water reabsorption results in increased BP.

Low blood pressure stimulates renin secretion from the kidney. Renin stimulates the production of angiotensin, which is converted to angiotensin II, which in turn stimulates aldosterone secretion from the adrenal cortex. Aldosterone stimulates Na⁺ ion and water reabsorption in the kidney.

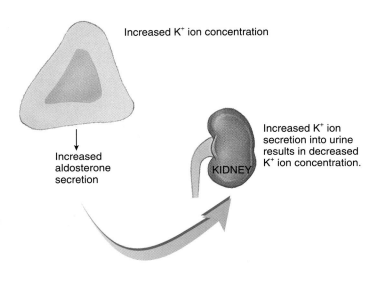

Increased K⁺ ion concentration

Increased aldosterone secretion

KIDNEY

Increased K⁺ ion secretion into urine results in decreased K⁺ ion concentration.

Elevated blood levels of K⁺ ions stimulate aldosterone secretion by the adrenal cortex. Aldosterone in turn causes an increased rate of K⁺ ion secretion by the kidneys.

Regulation of Sodium Ion Levels in Extracellular Fluids
Figure 27.2

An equilibrium exists between carbon dioxide, carbonic acid, and bicarbonate and hydrogen ions within erythrocytes. In erythrocytes, carbon dioxide reacts with water to form carbonic acid. An enzyme, carbonic anhydrase, catalyzes the reaction. Carbonic acid then dissociates to form hydrogen ions and bicarbonate ions. An equilibrium exists so that an increase in carbon dioxide causes more carbonic acid formation and a decrease in carbon dioxide causes some of the carbonic acid to dissociate to form carbon dioxide and water.

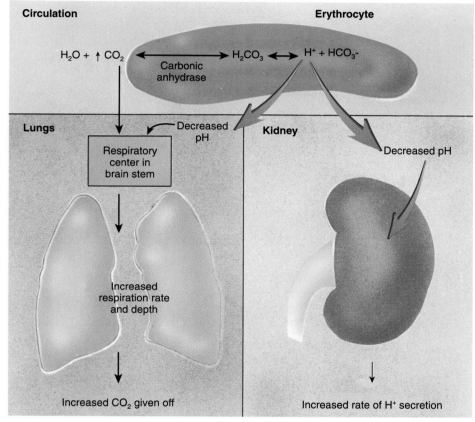

Circulation　　　　　　　　　　　　　　　**Erythrocyte**

$$H_2O + \uparrow CO_2 \longleftrightarrow H_2CO_3 \longleftrightarrow H^+ + HCO_3^-$$

Carbonic anhydrase

Lungs　　　　Decreased pH　　　**Kidney**

Respiratory center in brain stem

Decreased pH

Increased respiration rate and depth

Increased CO_2 given off

Increased rate of H^+ secretion

A decreased pH and increased PCO_2 stimulate the respiratory center and increase the rate of respiration resulting in an increased rate of carbon dioxide elimination.

A decrease in pH increases the rate of hydrogen ion secretion into the filtrate in the kidney.

Carbon Dioxide, Carbonic Acid, and Bicarbonate and Hydrogen Ions
Figure 27.7

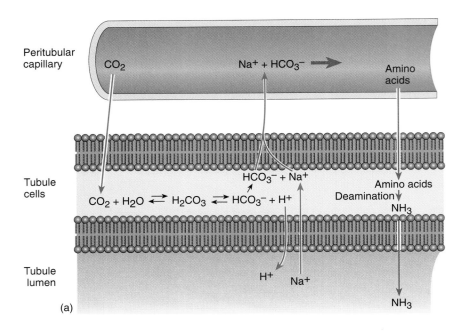

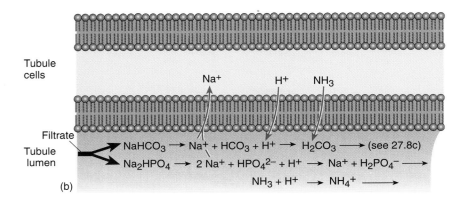

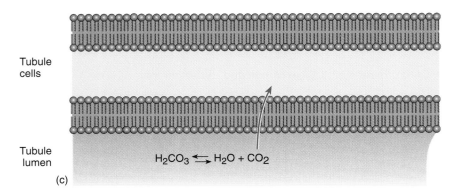

Kidney Regulation of Body Fluid pH
Figure 27.8

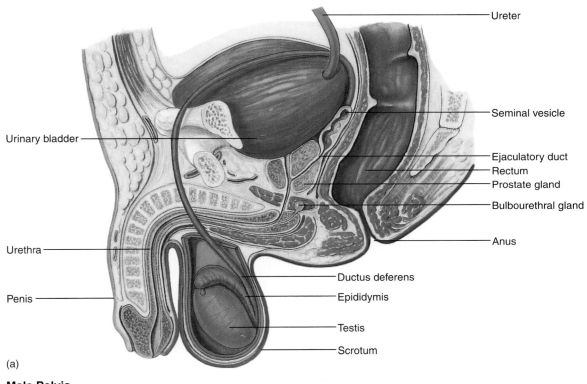

Ureter

Seminal vesicle

Urinary bladder

Ejaculatory duct

Rectum

Prostate gland

Bulbourethral gland

Anus

Urethra

Ductus deferens

Penis

Epididymis

Testis

Scrotum

(a)

Male Pelvis
Figure 28.1a

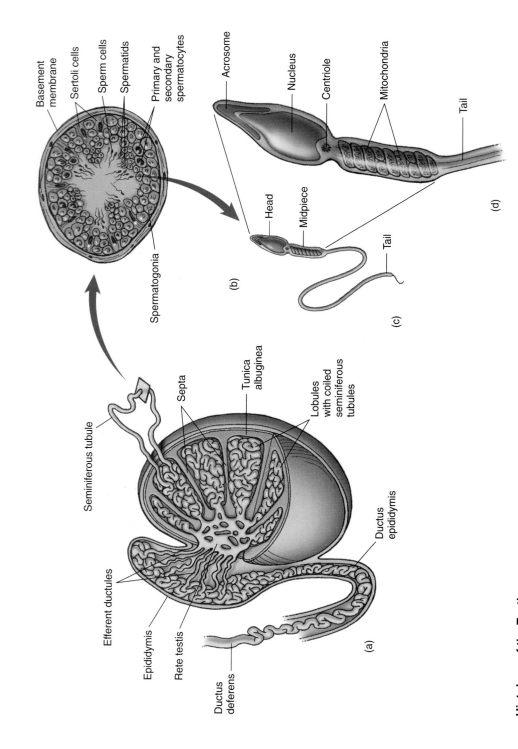

Basement membrane
Sertoli cells
Sperm cells
Spermatids
Primary and secondary spermatocytes

Spermatogonia

(b)

Acrosome
Nucleus
Centriole
Mitochondria
Tail

Head
Midpiece

Tail

(c)

(d)

Seminiferous tubule

Septa

Tunica albuginea

Lobules with coiled seminiferous tubules

Efferent ductules

Epididymis

Rete testis

Ductus epididymis

Ductus deferens

(a)

Histology of the Testis
Figure 28.2a–d

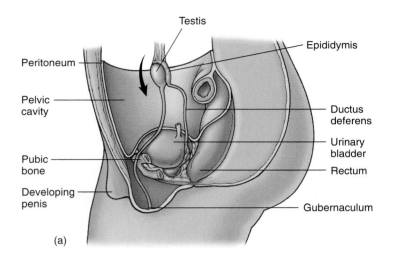

(a)

Labels for (a):
- Testis
- Peritoneum
- Pelvic cavity
- Pubic bone
- Developing penis
- Epididymis
- Ductus deferens
- Urinary bladder
- Rectum
- Gubernaculum

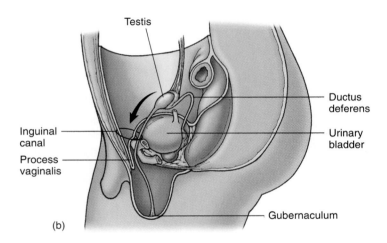

(b)

Labels for (b):
- Testis
- Inguinal canal
- Process vaginalis
- Ductus deferens
- Urinary bladder
- Gubernaculum

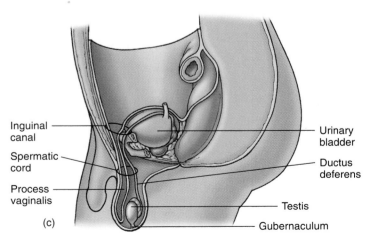

(c)

Labels for (c):
- Inguinal canal
- Spermatic cord
- Process vaginalis
- Urinary bladder
- Ductus deferens
- Testis
- Gubernaculum

Descent of the Testis
Figure 28.3

Mitotic division

Daughter cell

① Spermatogonia (germ cells)

46

46

46

First meiotic division

Primary spermatocyte

46

②

Secondary spermatocytes

23

23

Second meiotic division

③ Spermatids

23

23

23

23

④ Spermatids becoming sperm cells

23

23

23

23

23

23

23

23

23

Sperm cells

Lumen of seminiterous tubule

Basement membrane

Sertoli cell

Tight junction between Sertoli cells

Sertoli cell nucleus

Tight junction

Sperm cells

1. Spermatogonia are the cells from which sperm cells arise. The spermatogonia divide by mitosis. One daughter cell remains a spermatogonium that can divide again by mitosis. One daughter cell becomes a primary spermatocyte.

2. The primary spermatocyte divides by meiosis to form secondary spermatocytes.

3. The secondary spermatocytes divide by meiosis to form spermatids.

4. The spermatids differentiate to form cells.

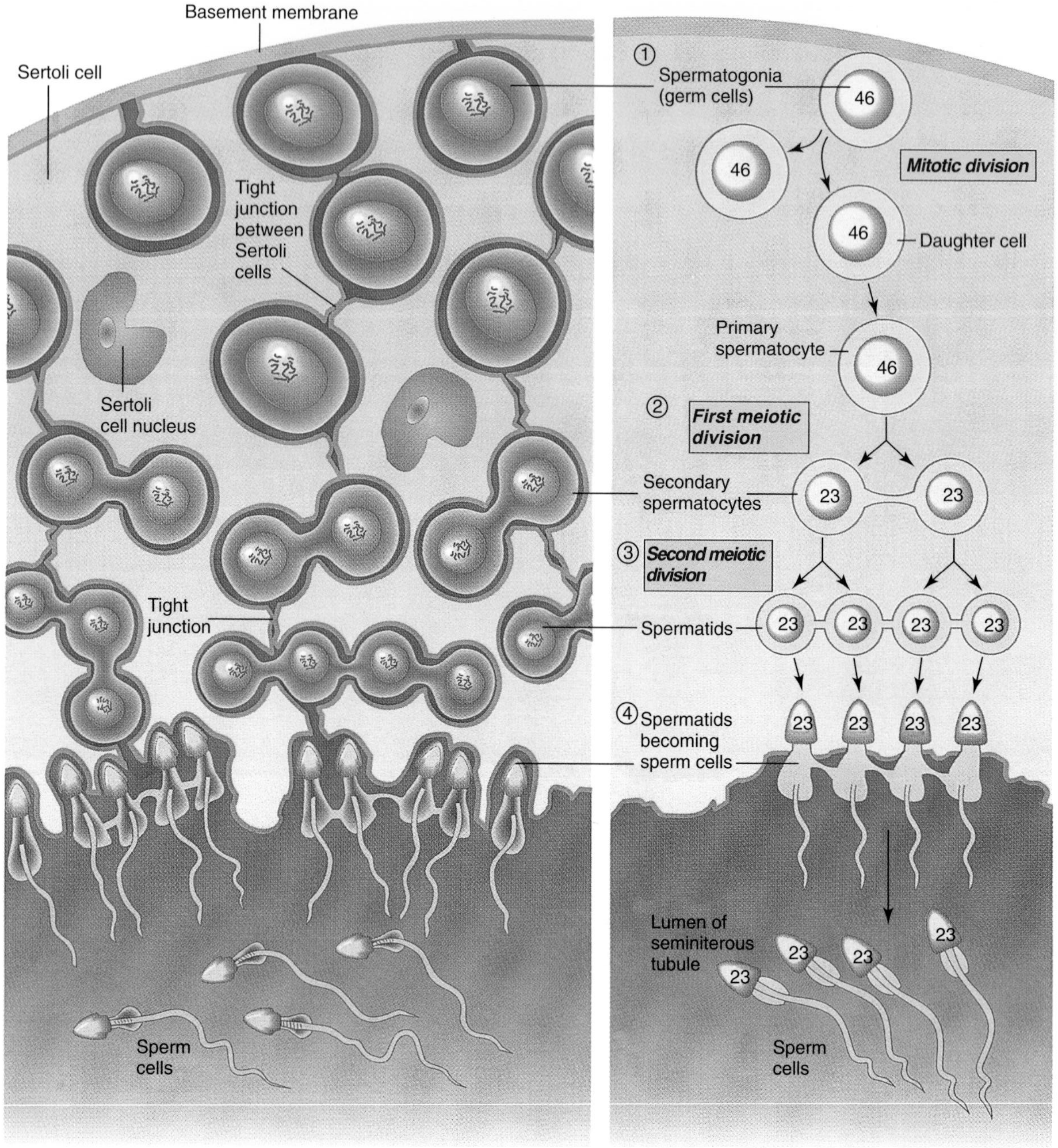

Spermatogenesis
Figure 28.4

322

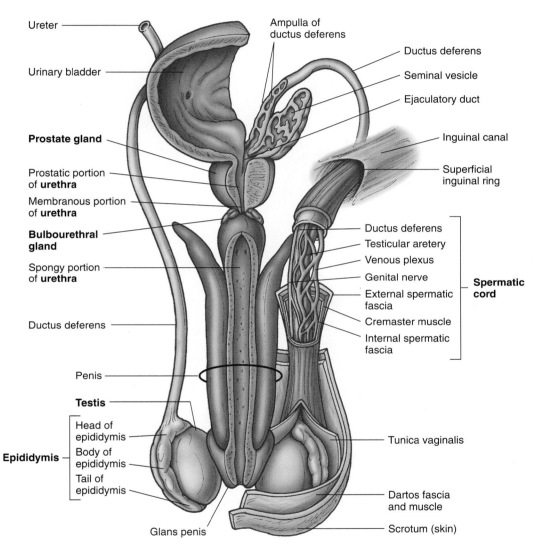

Frontal View of the Testis, Epididymis, Ductus Deferens, and Glands of the Male Reproductive System
Figure 28.5

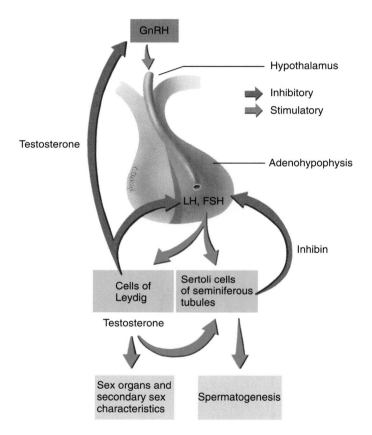

Regulation of Reproductive Hormone Secretion in Males
Figure 28.7

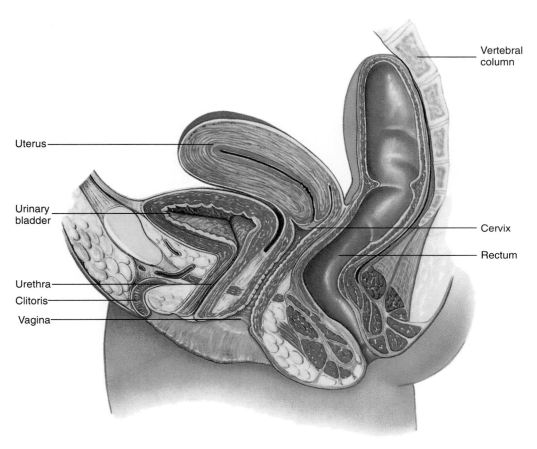

Uterus

Urinary bladder

Urethra

Clitoris

Vagina

Vertebral column

Cervix

Rectum

Sagittal Section of the Female Pelvis
Figure 28.8

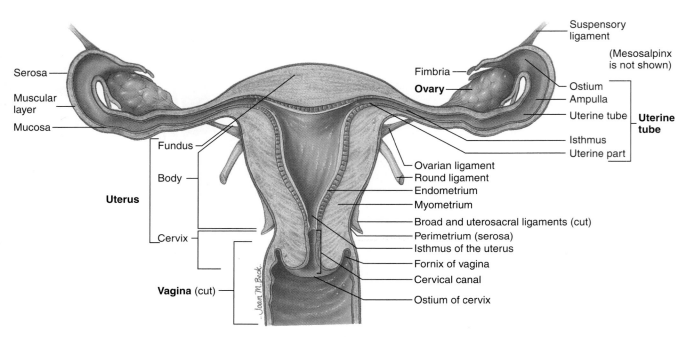

Serosa

Muscular layer

Mucosa

Suspensory ligament

(Mesosalpinx is not shown)

Fimbria

Ovary

Ostium

Ampulla

Uterine tube

Uterine tube

Isthmus

Uterine part

Fundus

Body

Uterus

Cervix

Ovarian ligament

Round ligament

Endometrium

Myometrium

Broad and uterosacral ligaments (cut)

Perimetrium (serosa)

Isthmus of the uterus

Fornix of vagina

Cervical canal

Ostium of cervix

Vagina (cut)

Joan M. Beck.

Uterus, Vagina, Uterine Tubes, Ovaries, and Supporting Ligaments
Figure 28.9

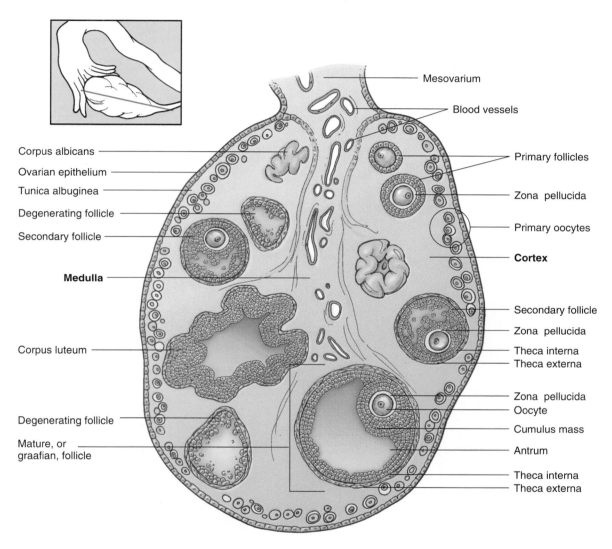

Corpus albicans

Ovarian epithelium

Tunica albuginea

Degenerating follicle

Secondary follicle

Medulla

Corpus luteum

Degenerating follicle

Mature, or
graafian, follicle

Mesovarium

Blood vessels

Primary follicles

Zona pellucida

Primary oocytes

Cortex

Secondary follicle

Zona pellucida

Theca interna
Theca externa

Zona pellucida
Oocyte

Cumulus mass

Antrum

Theca interna
Theca externa

Histology of the Ovary
Figure 28.10

1. The primordial follicle consists of an oocyte surrounded by a single layer of squamous granulosa cells.

2. A primordial follicle becomes a primary follicle as the granulosa cells become enlarged and cuboidal.

3. The primary follicle enlarges. Granulosa cells form more than one layer of cells.

4. A secondary follicle forms when fluid-filled spaces develop among the granulosa cells and a well developed theca becomes apparent around the granulosa cells.

5. A graafian follicle forms when the fluid-filled spaces form a single antrum. When a graafian follicle becomes fully mature, it is enlarged to its maximum size, a large antrum is present, and the oocyte is located in the cumulus mass.

6. During ovulation the oocyte is released from the follicle, along with some surrounding granulosa cells of the cumulus mass called the corona radiata.

7. Following ovulation, the granulosa cells divide rapidly and enlarge to form the corpus luteum.

8. When the corpus luteum degenerates, it forms the corpus albicans.

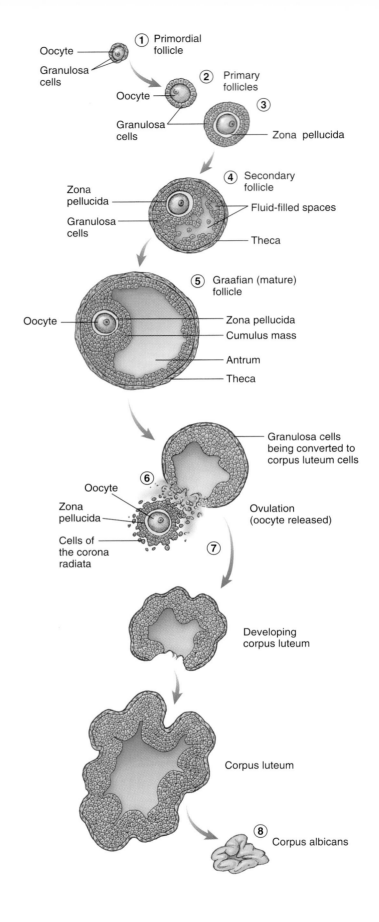

Maturation of the Follicle and Oocyte
Figure 28.11

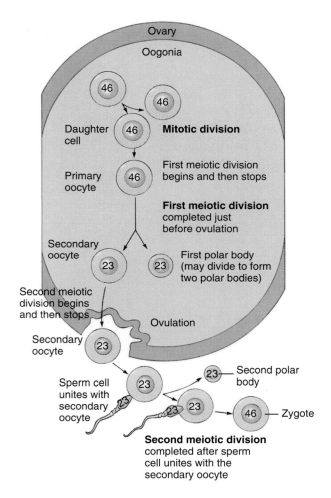

Maturation and Fertilization of the Oocyte
Figure 28.12

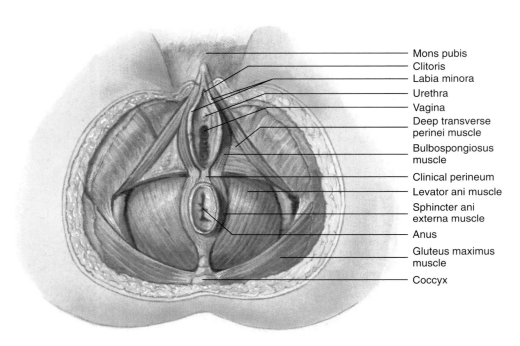

Mons pubis
Clitoris
Labia minora
Urethra
Vagina
Deep transverse perinei muscle
Bulbospongiosus muscle
Clinical perineum
Levator ani muscle
Sphincter ani externa muscle
Anus
Gluteus maximus muscle
Coccyx

Inferior View of the Female Perineum
Figure 28.14

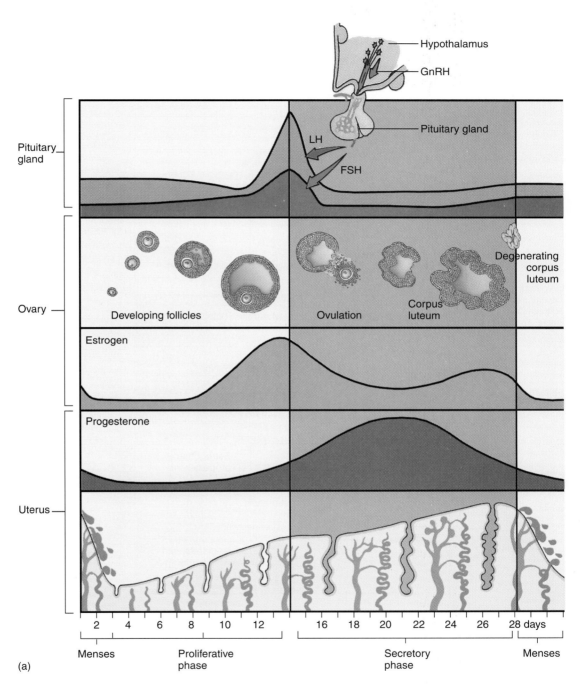

The Menstrual Cycle
Figure 28.16a

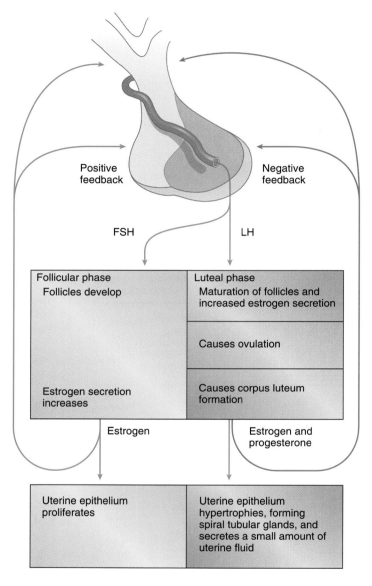

Positive feedback

Negative feedback

FSH

LH

Follicular phase	Luteal phase
Follicles develop	Maturation of follicles and increased estrogen secretion
	Causes ovulation
Estrogen secretion increases	Causes corpus luteum formation

Estrogen

Estrogen and progesterone

Uterine epithelium proliferates	Uterine epithelium hypertrophies, forming spiral tubular glands, and secretes a small amount of uterine fluid

(b)

Regulation of Hormone Secretion During the Menstrual Cycle
Figure 28.16b

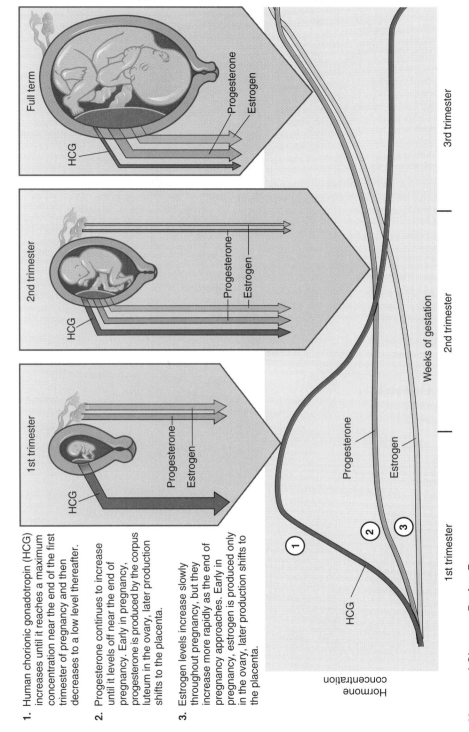

Hormonal Changes During Pregnancy
Figure 28.18

1. Human chorionic gonadotropin (HCG) increases until it reaches a maximum concentration near the end of the first trimester of pregnancy and then decreases to a low level thereafter.

2. Progesterone continues to increase until it levels off near the end of pregnancy. Early in pregnancy, progesterone is produced by the corpus luteum in the ovary, later production shifts to the placenta.

3. Estrogen levels increase slowly throughout pregnancy, but they increase more rapidly as the end of pregnancy approaches. Early in pregnancy, estrogen is produced only in the ovary, later production shifts to the placenta.

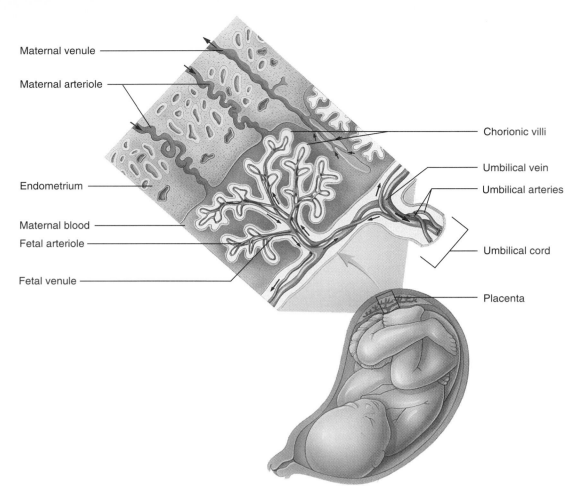

Maternal venule

Maternal arteriole

Chorionic villi

Endometrium

Umbilical vein

Umbilical arteries

Maternal blood

Fetal arteriole

Umbilical cord

Fetal venule

Placenta

Mature Placenta and Fetus
Figure 29.4

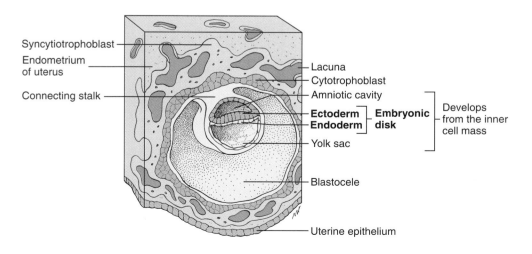

Syncytiotrophoblast

Endometrium
of uterus

Lacuna

Cytotrophoblast

Connecting stalk

Amniotic cavity

Ectoderm | **Embryonic**
Endoderm | **disk**

Develops
from the inner
cell mass

Yolk sac

Blastocele

Uterine epithelium

Embryonic Disk
Figure 29.5

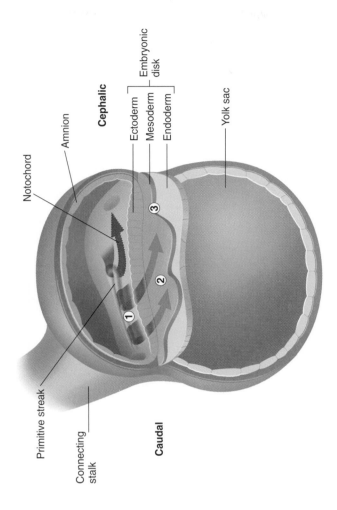

Notochord

Amnion

Cephalic

Ectoderm ⎤
        ⎬ Embryonic
Mesoderm ⎦ disk

Endoderm

Yolk sac

Primitive streak

Connecting stalk

Caudal

1. Cells in the surface ectoderm move toward the primitive streak and fold into the streak (*blue arrow tails*).

2. Cells that enter the primitive streak come out the other side of the streak as mesodermal cells (*red arrows*).

3. The mesoderm (*red*) lies between the ectoderm (*blue*) and endoderm (*yellow*).

Primitive Streak
Figure 29.6

1. The neural plate is formed from ectoderm.
2. Neural folds form as parallel ridges along the embryo.
3. Neural crest cells break away from the crest of the neural folds.
4. The neural folds meet at the midline to form the neural tube.

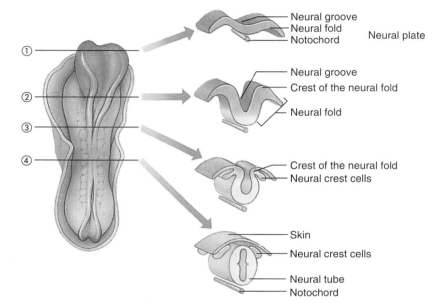

Neural groove
Neural fold
Notochord
Neural plate

Neural groove
Crest of the neural fold
Neural fold

Crest of the neural fold
Neural crest cells

Skin
Neural crest cells
Neural tube
Notochord

Formation of the Neural Tube
Figure 29.7

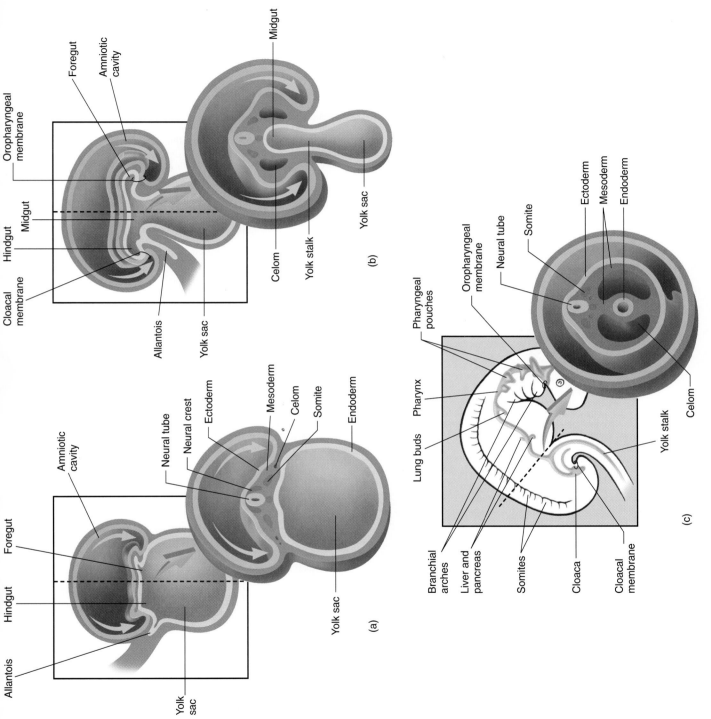

Formation of the Digestive Tract
Figure 29.8

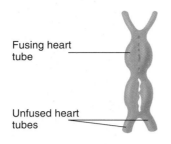

Fusing heart tube

Unfused heart tubes

(a) 20 days after fertilization. At this age, the heart consists of two parallel tubes.

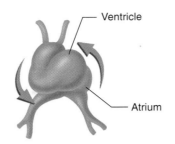

Ventricle

Atrium

(b) 22 days after fertilization. Fused, bent heart tube (*blue arrows suggest the direction of bending*) results from the elongation of the heart within the confined space of the pericardium.

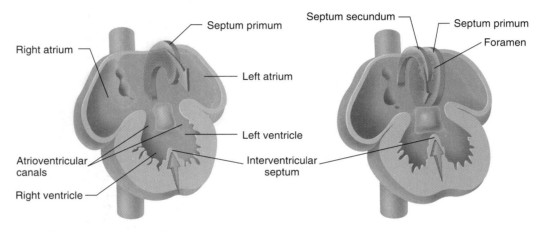

Septum primum

Right atrium

Left atrium

Left ventricle

Atrioventricular canals

Interventricular septum

Right ventricle

(c) 31 days after fertilization. The septum primum of the interatrial septum and the interventricular septum grow toward the center of the heart.

Septum secundum

Septum primum

Foramen

(d) 35 days after fertilization. The septum primum is complete and a foramen opens in the septum. The interventricular septum is nearly complete.

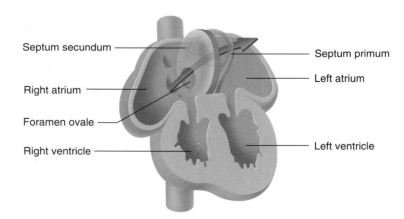

Septum secundum

Right atrium

Foramen ovale

Right ventricle

Septum primum

Left atrium

Left ventricle

(e) The final embryonic condition of the interatrial septum. Blood from the right atrium can flow through the foramen ovale into the left atrium. As blood begins to flow in the other direction, the septum primum is forced against the septum secundum, closing the foramen ovale.

Development of the Heart
Figure 29.11

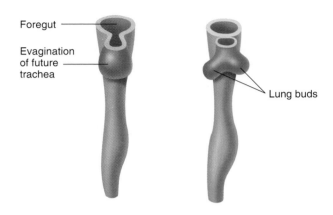

Foregut

Evagination
of future
trachea

Lung buds

(a) 28 days after fertilization. A single bud forms and divides
into two buds, which will become the lungs and primary
bronchi.

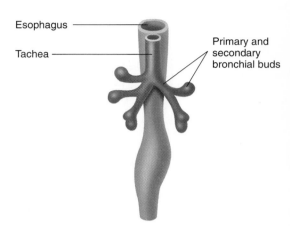

Esophagus

Tachea

Primary and
secondary
bronchial buds

(b) 32 days after fertilization. Primary bronchi branch to form
secondary bronchi, which supply the lobes.

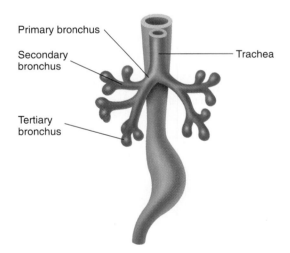

Primary bronchus

Secondary
bronchus

Trachea

Tertiary
bronchus

(c) 35 days after fertilization. Secondary bronchi branch to
form tertiary bronchi, which supply the bronchopulmonary
segments.

Development of the Lung
Figure 29.12

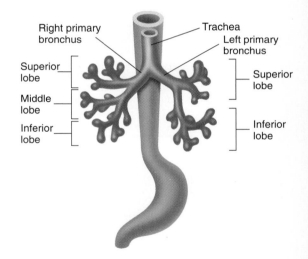

Right primary
bronchus

Trachea

Left primary
bronchus

Superior
lobe

Superior
lobe

Middle
lobe

Inferior
lobe

Inferior
lobe

(d) 50 days after fertilization. Continued branching.

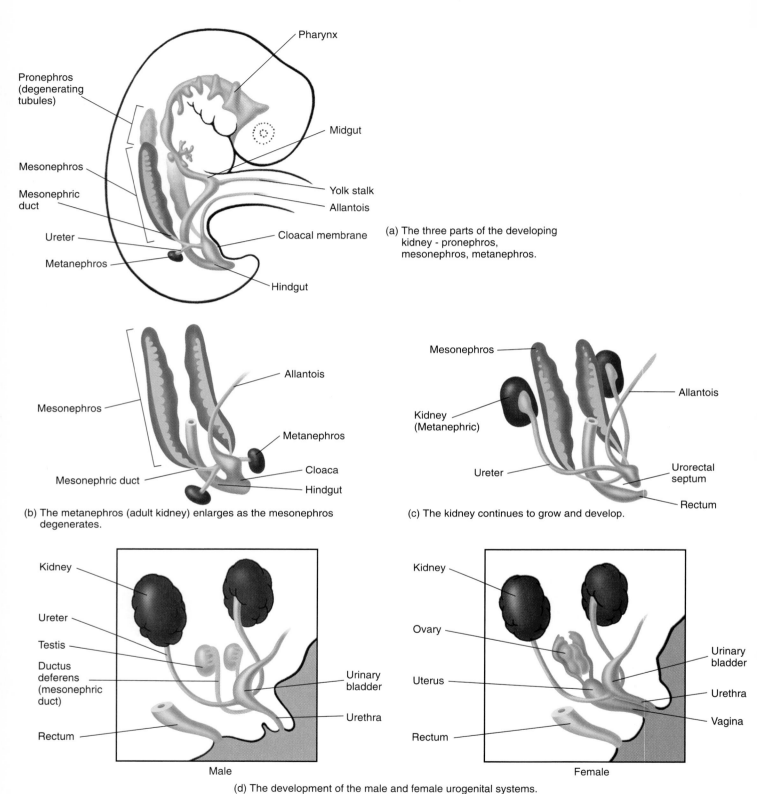

(a) The three parts of the developing kidney - pronephros, mesonephros, metanephros.

(b) The metanephros (adult kidney) enlarges as the mesonephros degenerates.

(c) The kidney continues to grow and develop.

(d) The development of the male and female urogenital systems.

Development of the Kidney and Urinary Bladder
Figure 29.13

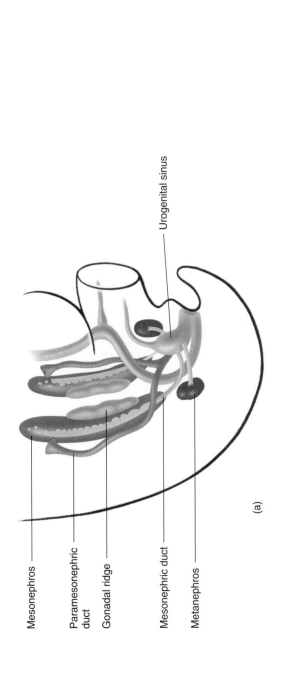

Mesonephros

Paramesonephric duct

Gonadal ridge

Mesonephric duct

Metanephros

Urogenital sinus

(a)

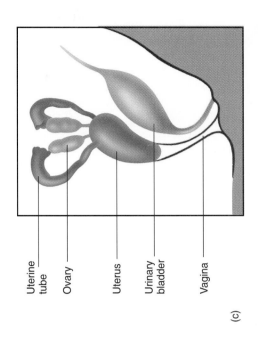

Uterine tube

Ovary

Uterus

Urinary bladder

Vagina

(c)

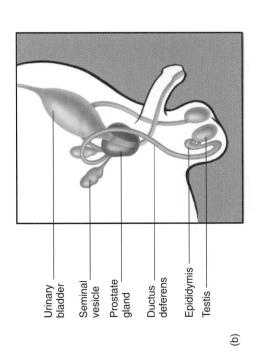

Urinary bladder

Seminal vesicle

Prostate gland

Ductus deferens

Epididymis

Testis

(b)

Development of the Reproductive System
Figure 29.14

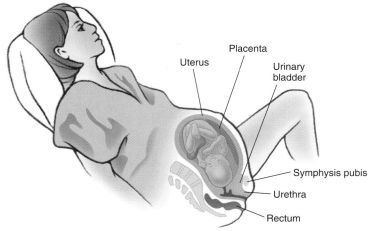

(a) The position of the fetus before parturition.

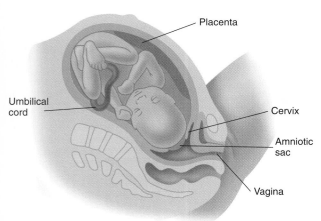

(b) The cervix begins to dilate.

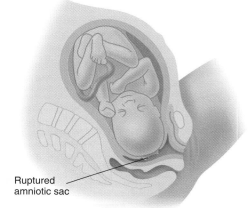

(c) Further dilation of the cervix and rupture of the amniotic sac occur.

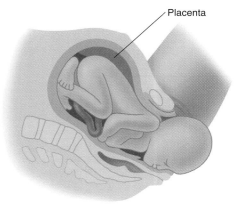

(d) The fetus is expelled from the uterus.

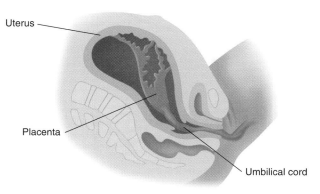

(e) The placenta is then expelled.

Process of Parturition
Figure 29.17

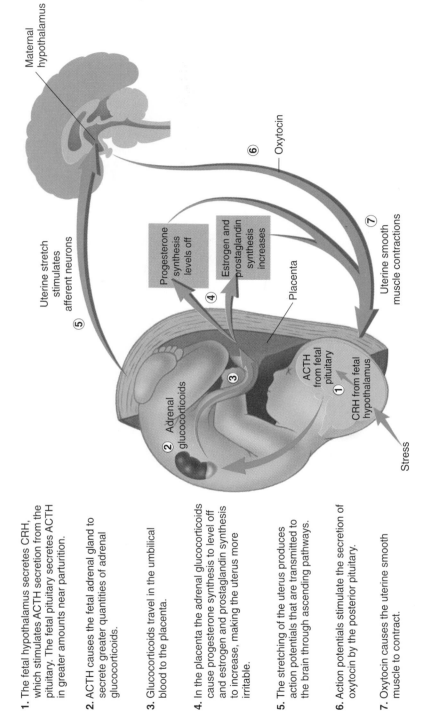

1. The fetal hypothalamus secretes CRH, which stimulates ACTH secretion from the pituitary. The fetal pituitary secretes ACTH in greater amounts near parturition.

2. ACTH causes the fetal adrenal gland to secrete greater quantities of adrenal glucocorticoids.

3. Glucocorticoids travel in the umbilical blood to the placenta.

4. In the placenta the adrenal glucocorticoids cause progesterone synthesis to level off and estrogen and prostaglandin synthesis to increase, making the uterus more irritable.

5. The stretching of the uterus produces action potentials that are transmitted to the brain through ascending pathways.

6. Action potentials stimulate the secretion of oxytocin by the posterior pituitary.

7. Oxytocin causes the uterine smooth muscle to contract.

Factors That Influence the Process of Parturition
Figure 29.18

343

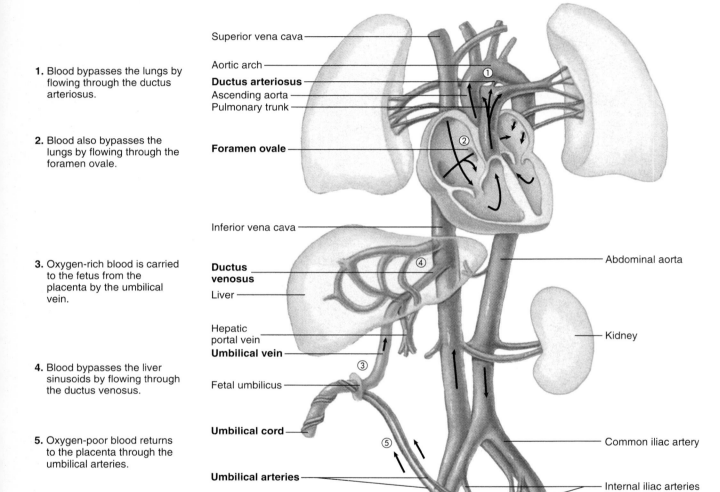

1. Blood bypasses the lungs by flowing through the ductus arteriosus.

2. Blood also bypasses the lungs by flowing through the foramen ovale.

3. Oxygen-rich blood is carried to the fetus from the placenta by the umbilical vein.

4. Blood bypasses the liver sinusoids by flowing through the ductus venosus.

5. Oxygen-poor blood returns to the placenta through the umbilical arteries.

Superior vena cava

Aortic arch

Ductus arteriosus

Ascending aorta

Pulmonary trunk

Foramen ovale

Inferior vena cava

Ductus venosus

Liver

Hepatic portal vein

Umbilical vein

Fetal umbilicus

Umbilical cord

Umbilical arteries

Abdominal aorta

Kidney

Common iliac artery

Internal iliac arteries

(a)

Circulatory Changes at Birth
Figure 29.19 (Continued)

1. When air enters the lungs, the ductus arteriosus closes and becomes the ligamentum arteriosum.

2. The foramen ovale closes and becomes the fossa ovalis.

3. The umbilical arteries and vein are cut so that the umbilical vein becomes the ligamentum teres (round ligament) of the liver.

4. The ductus venosus becomes the ligamentum venosum.

5. The umbilical arteries become the umbilical ligaments.

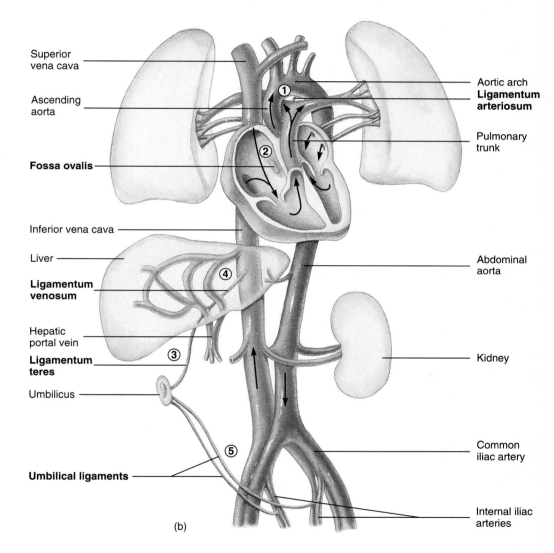

Superior vena cava

Ascending aorta

Fossa ovalis

Inferior vena cava

Liver

Ligamentum venosum

Hepatic portal vein

Ligamentum teres

Umbilicus

Umbilical ligaments

Aortic arch
Ligamentum arteriosum

Pulmonary trunk

Abdominal aorta

Kidney

Common iliac artery

Internal iliac arteries

(b)

Figure 29.19

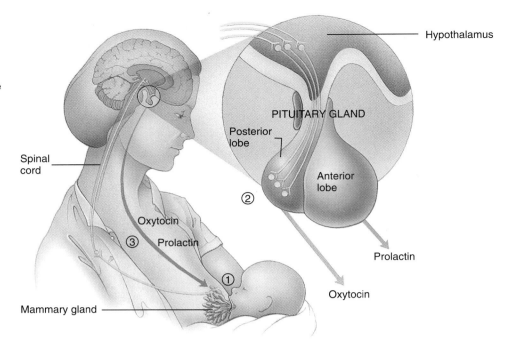

1. Stimulation of the nipple by the baby's suckling initiates action potentials in the afferent neurons that connect with the hypothalamus.

2. The hypothalamus stimulates the posterior pituitary to release oxytocin and the anterior pituitary to release prolactin.

3. Oxytocin stimulates milk release from the breast. Prolactin stimulates additional milk production.

Hormonal Control of Lactation
Figure 29.20

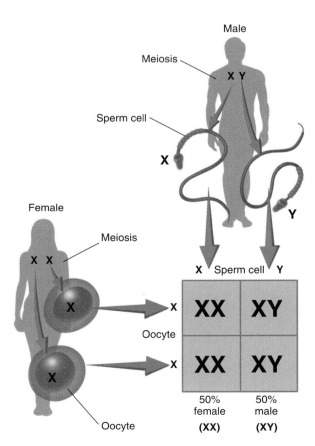

Examples of Inheritance Patterns
Figure 29.23

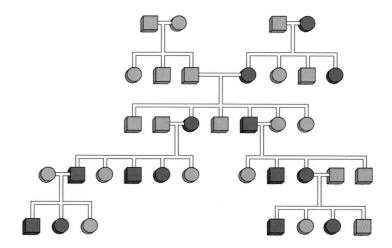

Pedigree of a Simple Dominant Trait
Figure 29.27

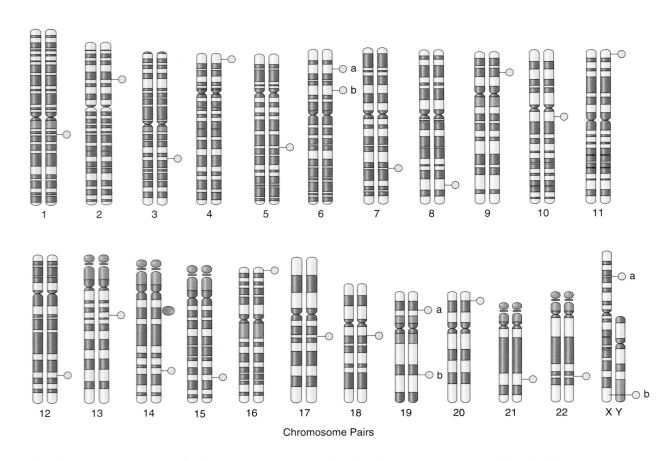

Chromosome Pairs

1. Gaucher disease
2. Familial colon cancer*
3. Retinitis pigmentosa*
4. Huntington disease
5. Familial polyposis of the colon
6a. Spinocerebellar ataxia
6b. Hemochromatosis

7. Cystic fibrosis
8. Multiple exostoses*
9. Malignant melanoma
10. Multiple endocrine neoplasia, type 2
11. Sickle cell disease
12. PKU (phenylketonuria)

13. Retinoblastoma
14. Alzheimer's disease*
15. Tay-Sachs disease
16. Polycystic kidney disease
17. Breast cancer*
18. Amyloidosis
19a. Familial hypercholesterolemia
19b. Myotonic dystrophy

20. ADA deficiency
21. Amyotrophic lateral sclerosis*
22. Neurofibromatosis, type 2
Xa. Muscular dystrophy
Xb. Factor VIII deficiency (hemophilia A)

*Gene responsible for only some cases.

The Human Genomic Map
Figure 29.29